Simple Signature Dishes

推薦序 Forewords

鼎爺

精雕細煮！祝新書大賣！

林峯

Bonnie 是我認識了十多年的好友，我們非常投契，在人前人後我亦稱她為「家姐」。

「家姐」才藝雙全之餘，對朋友非常之好而且為人幽默，絕對是朋友圈中的開心果！

我有幸在多次的朋友聚會中嚐過她親手所煮的不少好餸！知道這次她再次出書分享多年煮食心得，真心推介給各位朋友！阿家姐，何時有空再煮妳的拿手小菜給我品嚐呀！？

黃宗澤

某年某月某日，心思思想煮鹽焗蟹，工欲善其事，先上網做一下資料搜集，結果彈出一網誌，名為「玻璃朱城」，點進細閱，讀到文末發現一張照片，真是廣告有云～個心都離一離！原來玻璃朱……Bonnie Chu 是也！朱小姐……

多年不見的一位朋友，還記得此女子從事金融業，跑到網上寫美食，原因十居其九是……錢多，錢多者嗜好多亦是自然。但萬想不到近年朱小姐居然在電視上露面，當上了烹飪節目主持，頭頭是道。常言道，人是需要有冒險精神的，但冒險到拿自己一半的事業來投資，而且是「錢」途無可限量的金融業，這不叫投資，這叫投河自盡，可想而知，朱小姐絕對不是開玩笑，她是拿着自己的事業當本錢，寫着一本她對美食的血淚史。小弟愚見，這樣的作者寫的食譜，誠意、態度、性價比……足矣！

張柏芝

已經忘了認識 Bonnie 有多少年，但跟她熟悉之後才知道她擁有天使的面孔及魔鬼的身材，常常在夏天的時候穿著非常性感的比堅尼。如果我不是比她漂亮 1000 倍的話我一定妒忌她。

和大家簡單介紹，她是一個辣媽，一個追求時尚潮流的一個女性，她極度喜歡波鞋，是瘋癲的；她是一個很盡責任的好媽媽，但最終想和大家説，沒想到她更是一個出得廚房、入得廳堂的完美女人，她煮的美食是從內心裏面的靈魂和愛煮給大家，極度推介她所有的食譜和她教學的課程，不管是教成人或是教小朋友，她同樣地付出 100 個誠意，一句到尾，正到爆、索到爆、好食到爆！

陳凱琳

I remember the first time I was introduced to Bonnie Chu. It was in July 2017, and immediately upon meeting her, she asked me if I wanted to learn how to cook. Even though I'm not a regular in the kitchen, cooking was always something I was secretly interested in. So when Bonnie invited me to her newly opened studio to join her classes, I really felt like it was a dream come true, and so began my love for cooking!

Over the years, Bonnie and I became good friends through our mutual love for food. I recall the time she gifted me her first cookbook. After ripping through the pages, I told her that I wanted to learn as many recipes as I could! Interestingly, Bonnie's cookbook infused both traditional Chinese dishes and also authentic Thai dishes that she learned while studying at Blue Elephant in Bangkok. Every recipe in the book was tried and perfected by the cook herself, but what's more was that they weren't just instructions on a page, but words of warmth and passion! In a way, when I read the cookbook, it was as if Bonnie was standing in front of me, instructing me on "how to marinate the chicken, save the chicken fat and use it later in the rice to make it fragrant and delicious!"

In the beginning of 2019, Bonnie messaged me and shared with me that she was getting ready for her second cookbook. At that moment, all I could think of was, about time! I can't express how happy I am for her and how excited I am to get my hands on her new publication. It looks like Bonnie will be seeing a lot of me this year once her cookbook comes out!

龔嘉欣

從讀書時候開始，我一向很喜歡在網上看關於煮食甜品的資訊，那時我記得印象很深刻的一個名字——玻璃朱。當時不明白她的名字由來，就是覺得特別，從那時就開始默默追蹤她，會翻看她之前的食譜，也會期待着她即將推出的食譜，為甚麼我那麼喜歡她的食譜？因為她總會把看似複雜的食譜簡單化，讓我們每一個人都能容易地煮出一手好菜。相信很多人都試過，跟着食譜，但煮出來相差很遠，甚至感到氣餒。謝謝 Bonnie 的食譜，每一次我煮給朋友們，看到他們吃得開心滿足，我也會有一份成功感。

到後來，我們成為了朋友，除了煮食上她會不厭其煩的教我，把所有步驟錄好音給我，怕我忘記，非常細心。於朋友，她總會義無反顧，遇到甚麼需要幫忙的，她一定在你身邊，而且我也不明白，為甚麼她甚麼都懂，想到甚麼問她，她都一定解答到，我想她應該是一個 superwoman ！

姚子羚

認識這位好友雖然不太久～但這位美女廚神玻璃朱，除了廚藝了得亦是一位義氣朋友！她知道我吃素……還特意在她的廚藝班教了我做我至愛的水煮齋！易做又好味～家常小炒到烹調複雜的菜式也難不倒她……全因她虛心學習，向不同師傅請教再用心鑽研～這位好姊妹的第一本烹飪書我當然有好好保存！這次再推出，我非常期待～也預祝妳新書大賣！姊妹們一定撐撐撐～

羅霖

認識 Bonnie 的朋友都知道她是個很爽直的人，所以她煮的餸菜都是「快・靚・正」，這個亦是我最欣賞的地方。因為我們這種新時代女性，要兼顧家庭和事業，所以做菜也不能像以往般那樣精雕細琢。用最簡單的方法做出最精美的菜式是一個新趨勢。而 Bonnie 不止是中西餐也煮得好、連東南亞美食也煮得十分出色。今次她再出這新食譜我一定大力推薦！因為上一次我買了很多她的食譜送給朋友，朋友們都反映説跟着她的做法煮出來的味道真的十分靠譜。所以希望大家也可以支持一下～

Jacky Yu

在烹飪的領域，高手隨時都可能是你身邊的其中一個朋友，自從看到 Bonnie 第一本的處女作《星級菜自家煮》之後，我就覺得 Bonnie 就是一個不折不扣的「女煮人」，懂得煮之外，還非常懂得如何將自己的菜式與烹飪愛好者分享。

現在的 Bonnie，更是一位非常多元化的烹飪達人，除了不時在自己的社交平台分享她的美食之外，還會參與不同的烹飪節目和美食活動。我最欣賞的是她能非常清晰、直接簡易地將她的食譜以最佳的效果烹調出來，而且她都非常大方，毫不吝嗇分享食譜，令學習者真的可以將菜式在家的餐桌上完美展現，派上用場。

Bonnie 的最新食譜著作《星級菜簡單煮》同樣貫徹她的簡易作風，將在街外餐廳吃到的大菜，以家庭主婦的簡單技巧，在家都可以輕易烹調出來，而且保證百分百好食，色香味美！

期待 Bonnie 的最新力作，祝大賣！

林盛斌 Bob

喺香港，專業嘅女廚師其實唔算好多，堅係有料兼煮得好食嘅就更加少，你話仲要又好樣又索又好身材嘅，好似真係得番 Bonnie 一個⋯⋯

P.S. 以上內容，係我俾 Bonnie 用個鑊鏟指住個頭嘅狀態下寫嘅！

麥美恩

Food is the best ice breaker. It's the reason how I met Bonnie thru "我係小廚神 3". Congrats to the new cook book!

黃碧蓮

I worked with Chef Bonnie while shooting "我係小廚神 3". She is a passionate culinarian and possesses such natural instincts when it comes to food, definitely a privilege if you get to observe her but an honour if you get to taste her cooking!

自序 Preface by Bonnie

足足等了 4 年，我的第二本個人食譜書終於可以跟大家見面了。其實，當 2015 年完成了我的首個 BB 後，那刻已經覺得有點心力交瘁……哈哈！不是誇張，當時真的嘔心瀝血地完成了首本著作後，真的好想休息一下，跟自己説還是應該「貴精不貴多」，不要每年出書了。但 2016，還是被邀請與另外兩位煮食好友出版了一本合拍食譜，還好還好……只有 20 個食譜，製作亦比較簡單，勉強還應付到，但之後真的決定要停一停了。

而在這幾年期間也發生了好些人生大事，令到自己更分身不暇……首先經歷了離婚、跟着便埋頭苦幹開創了我的廚藝學校「Cooktown」！然後被 ViuTv 邀請拍攝煮食節目「煮餐飯有幾難」。之後又再被 TVB 邀請拍攝「我係小廚神 3」及加入了 Big Big Channel 做 KOL……最重要的是還要照顧我的寶貝兒子 Keon ！老實説真是忙得無暇再想出食譜書……但亦慶幸因為我有廚藝學校的關係，我的第一本著作出版幾年仍長賣長有，至今已經印至第四版，賣了接近一萬書了！以第一次出書來説已經算有不錯的成績。真的要衷心多謝大家一直以來的支持！感激感激。

2018 年中後，Cooktown 改變了經營模式，令我可以有更多空間去做自己的事。記得曾碰到位舊朋友，那時她剛加入了「萬里機構」出版社，亦曾提出可以一起合作出書事宜……而自己在這幾年間亦累積了更多經驗，想想也是時候出版第二本個人食譜了，何況還是跟香港最大的出版社合作，當然自覺十分榮幸及不想錯過這機會，結果便一拍即合了。

我永遠不敢稱呼自己為大廚，因為我並沒有在餐廳當過廚。我只是一位「家廚」！所以，我的食譜一向都貫徹我「化繁為簡」的方法，喜歡把飯店的菜式用個簡單些的方式去煮，但又不失餐廳口味。一眾在職女士、太太、媽媽們忙完一天的工作及照顧家庭，請問那還有時間在廚房「雕花」？當然，偶而做個繁複的菜式去挑戰一下自己當然很有樂趣；但要每天做的卻又不同了。然而，簡單並不代表是馬虎。做得出來一定是有水準，才過得自己啊！

開始公開寫食譜至今，真的聽過很多朋友讚我的食譜「真材實料」。説跟着煮真的可做出跟原本菜式十分相近的味道……這真是過獎了。我一向抱着「無私奉獻」的心態去分享食譜，絕不會留

COOKTOWN
COOK & STYLE

起一點點，怕別人青出於藍。很多人便因自私而令很多食譜或絕學失傳。我覺得別人若能把我的菜式發揚光大比只有我自己懂得煮會更有成功感。第二，就算我給你一樣的材料及煮法，但你也未必會煮得比我好，因為手勢、火喉和經驗亦絕對有影響！這在我教班時同學仔便常常說為甚麼老師煮的總是好吃點了……哈哈～

我還記得 7 歲時在母親節當天，用了十元（那些年一天零用只得 $2.5）買了兩塊厚火腿及蛋，還有些配菜煮給媽媽吃！那個算是我第一次煮即食麵以外的東西。火腿焦了一點、蛋也煎得不太好。記得媽咪也好像沒有吃完……這算是我這一生煮過最不滿意的一餐。之後，我便經常在廚房「搞搞震」、去上烹飪班、問我廚藝出眾的姨媽、買餸時又問街市佬、甚至直接問酒樓大廚及老闆；師傅鼎爺也是其中一個給我煩着的，哈哈……其實，每一件事上，只要抱着不斷學習及謙虛求教的精神，相信每個人也可以由地獄廚神變成為大廚的。尤其是做料理，全球多國菜式及上百萬種食材，你哪會有全部學懂的一天？到現在我仍然是不斷虛心學習中……

希望我在這書向大家分享的點滴，可以經你們再發揚光大；煮得更好，感染更多人喜愛下廚，把「愛」經由煮食去分享給家人及朋友。

最後，要感謝幫我拍攝食譜的團隊如 Humble Jim、阿權及化妝師 Meegan、Rachel。 還有 Keon 的 3 位契媽 Crystal，Carol 同 Grace 的仗義幫忙。一班贊助商們如 Fisher & Paykel 和 Le Creuset，使這本書才能順理完成。還有一班抽空為本書寫序的師傅、朋友……萬分感謝！

After 4 years of waiting, my second cookbook is finally ready. To be honest, when I completed the first creation in 2015, I felt exhausted...Haha! It was not an exaggeration. After completing the first book, I really wanted to take a break and told myself not to publish a book each year, to hold true to the principle of “it's quality rather than quantity that counts.

”That said, in 2016, I was invited to co-publish a cookbook with two other friends who are also into cooking. Fortunately, there were only 20 recipes. The production was relatively simple and I could still cope with it.

After that, I made up my mind to take a long break.

In the past few years, there have been many life changing events that took up all of my time and energy. The first was my divorce. After that I devoted all my time to establish my culinary school "Cooktown". Followed by ViuTv's invitation to shoot a cooking program "How Hard to Cook.", and then with TVB to shoot "Chef Minor 3". Thereafter I joined Big Big Channel as KOL. On top of all these, the most important responsibility is to take care of my son Keon. There was hardly any time left to think of a new cookbook. I am also thankful that because of the influence of my culinary school, my first cookbook is now at its 4th edition and sold close to 10,000 copies. As a first book, the response is considered pretty good. I am thankful and grateful for your continual support.

At mid-2018, "Cooktown" changed its business model. I was able to have more personal time to explore new ventures. At that time I met up with an old friend who has just joined Wan Li Book Co., Ltd . She brought up the idea of a new book. Having accumulated more experience these few years, I felt that it is the right time to come out with the second cookbook too, not to mention it will be under the umbrella of the biggest publisher in Hong Kong. I felt honoured and decided not to miss this opportunity.

I would not call myself a chef because I have never been the chef of any restaurant. I am just a “home cook”.

My recipes would always follow the principle of “Simplification”, using simpler methods to prepare the dishes served in restaurants without compromising the taste. Will working lady, wife or mother has the time to create beautiful garnishing in the kitchen after a hard day's work in the office or at home? Of course once in a while it can be quite fun to work on some complicated dishes as personal challenge. But it is another matter altogether if one has to do it every day. However, simplicity does not mean being sloppy. One should take pride to uphold certain standard.

From the time I started to write recipes for the public, I have received many praises about my recipes being “the real deal”, that if one follows the steps, the end product can actually taste very similar to that of the original dish. I am flattered. My sharing has always been selfless and nothing is held back. There are some

COOKTOWN

who prefer not to disclose everything leading to the loss in precious recipes or skills. I personally feel that if someone else is able to take my recipe to the next level, self satisfaction is much greater. Furthermore, when following the exact same ingredients and method, someone else' creation may not surpass mine due to the difference in skill set and experience. My culinary class students like to comment that the teacher's cooking always taste better!

I still remember that when I was 7 years old, on Mother's Day, I used ten Hong Kong dollars (those days the pocket money was only $2.5 a day) to buy two thick slices of ham, eggs and some side ingredients to cook for my mother. That was the first time I cooked something other than instant noodles. The ham was a little burned and the egg was not well cooked. I think my mum did not finish the food. That was the most unsatisfactory meal I have ever cooked in my life. After that, I would often dabble in the kitchen, attended cooking class, sought advice from my aunt who is an excellent cook, asked the hawkers in the wet market where I bought stuff from, even directly ask the chefs and owners of restaurants. Master Ding was one of those whom I went after. I believe that as long as one is willing to learn and possess a teachable attitude, it is definitely possible to transform from a hellish cook to a skilled chef. In the culinary world, there are multi-national cuisines and millions of ingredients, there is no way one can master everything. I will keep learning.

It is my sincere wish that the knowledge shared in this cookbook can be further expanded by the readers to create better dishes and to influence others to take up cooking, use cooking as a channel to spread Love to family and friends.

Last but not least, I want to express my sincere thanks to the team that helped me a lot with the photo shoot, such as Humble Jim, Kuen and makeup artist Meegan, Rachel; the good help from Keon's god mothers Crystal, Carol and Grace; the wonderful sponsors and support by Fisher & Paykel and Le Creuset; friends and teachers who wrote the forewords for this book. I could not have done it without all of you.

Bonnie

目錄 Contents

Chinese Delicacy

中式滋味

Asian Gourmet
亞洲風味

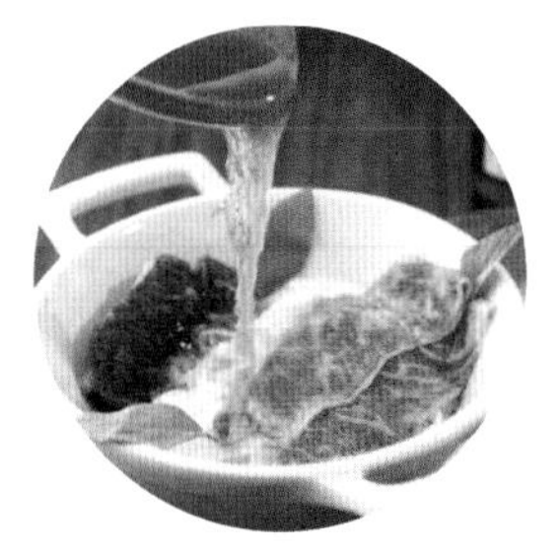

Western Delicacy

西式風味

COOKTOWN

Chinese Delicacy

中式滋味

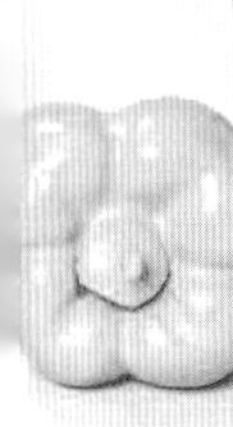

花雕雞油蛋白
烏冬蒸花蟹

Steamed Crab with Egg White and Udon in Hua Diao Wine and Chicken Fat

我自己最喜歡吃的就是「蟹」了。
蟹本身便很有鮮味，
不用多加調味也可以煮出美味的菜式，
以下便是其中一款。

Crab is my personal favourite.
As crab itself is very tasty,
little seasoning is required to turn it into a delicious dish.
This is one such recipe.

材料（2－4 人份量）

花蟹 2 隻（約 1 斤）
烏冬 1 個（解凍）
蛋白 3 隻
雞湯 100 毫升
雞油 2 湯匙

Ingredients (Serves 2-4)

2 flower crabs (about 1 catty)
1 pack udon (thawed)
3 egg whites
100ml chicken broth
2 tablespoons chicken fat

調味料

鹽半茶匙
糖 1 茶匙
魚露 1 茶匙
花雕 3 湯匙

Seasonings

1/2 teaspoon salt
1 teaspoon sugar
1 teaspoon fish sauce
3 tablespoons Hua Diao wine

隨個人喜好，最後可加些葱花。

Feel free to add some diced spring onion according to personal preference.

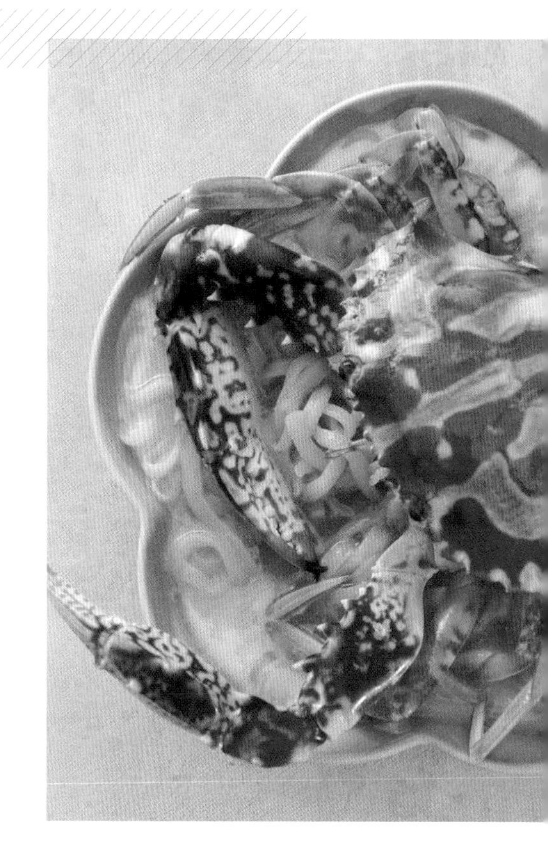

做法

1) 雞脂肪煎至出油，隔渣後，油備用。蟹斬件，備用。
2) 蛋白打勻，加入雞湯、鹽、糖、魚露攪勻。
3) 把烏冬放在碟底，加入蛋白汁，把蟹件鋪面，先加入 2 湯匙花雕及 1 湯匙雞油，大火蒸 10 分鐘。
4) 蒸熟後再多加入花雕及滾熱的雞油各 1 湯匙，即成。

Steps

1) Stir fry the chicken fat to release the oil. Strain and set aside the chicken oil. Chop the crabs into pieces.
2) Beat the egg whites. Add chicken broth, salt, sugar and fish sauce. Mix well.
3) Place the udon in plate. Add into the egg white mixture. Topped with the crab pieces. Add 2 tablespoons Hua Diao wine and 1 tablespoon chicken oil. Steam over high heat for 10 minutes.
4) Steam until done. Add 1 more tablespoon each of Hua Diao wine and boiling hot chicken oil. Serve.

可用陳村粉取代烏冬，或只用白蛋也可。

Udon is either optional (using only egg white) or can be replaced by Chen Chun noodle.

LE CREUSET

黃金蝦

Golden Prawns

平時出外用膳十分昂貴的菜式，在家既實惠又輕易便可以做到。這便是其中一道。

Some dishes are rather costly when served in restaurants but can be easily prepared at home without having to pay through the nose. This recipe is one such example.

材料（2-4 人份量）

中蝦 10 隻

鹹蛋 6 隻

牛油 20 克

調味料

糖 1 茶匙

Ingredients (Serves 2-4)

10 medium prawns

6 salted egg yolks

20g butter

Seasonings

1 teaspoon sugar

我個人喜歡煮蝦連殼的，覺得鮮味會更為突出。若你喜歡「啖啖肉」的可以先去殼開邊去腸、再沾粉走油。記着去殼時要保留蝦尾殼，炸起來賣相會更漂亮。

I like to cook prawns without removing the shells and it makes the flavour become more prominent. If you prefer the meaty mouth-feel, peel and slit prawns to remove the intestine, coat prawns with potato starch followed by deep-frying. Remember to retain the tail when removing the shell. The presentation will be more attractive.

做法

1) 鹹蛋先蒸 20 分鐘將蛋黃取出並壓爛成蓉，備用。
2) 中蝦洗淨及抹乾，剪去頭及腳，沾上生粉，走油炸 1、2 分鐘。拿起備用。
3) 中小火把牛油溶掉，加入蛋黃炒勻至略有泡沫，加入蝦炒勻，再加入 1 茶匙糖，把蛋黃炒至全掛在蝦身上，即成。

Steps

1) Steam the salted egg for 20 minutes. Remove the egg yolk and mash it very fine. Set aside.
2) Rinse and pat dry the prawns. Trim off the prawn head and legs. Coat the prawns with potato starch. Deep fry in hot oil for 1 to 2 minutes. Dish up.
3) Melt the butter over medium low heat. Add the mashed salted egg yolk and stir-fry until foamy. Add the prawns. Stir until even. Add 1 teaspoon sugar. Stir-fry until prawns are coated with mashed salted egg yolk. Serve.

貼士
tips #2

炸粉請用生粉，不要用粟粉。會更加脆身。

For crispier mouth-feel, use potato starch instead of corn starch.

Chinese Delicacy
中式滋味

芥菜膽大魚頭

Big Fish Head with Mustard Green

這道菜，可以是一道湯，又可以是一道菜。
而這道菜我在一家吃宵夜的店是必叫的，
在家做其實也是很容易的。

This dish can be regarded either as a soup or a dish.
It is a must-have for me
whenever I visit a restaurant that serves supper.
It can also be easily prepared at home.

材料（2-4 人份量）

大魚頭 1 個（約一斤、斬件）

芥菜膽 1 個（洗淨切塊）

草菇 8 粒（洗淨開邊）

硬豆腐 2 磚（切大塊）

薑片 4 大片

葱花 1 湯匙

調味料

胡椒粉 2 茶匙

鹽 1 茶匙

糖 1 茶匙

米酒 2 茶匙

做魚湯要奶白色，倒入的一定要滾水，再加大火大滾才行。

To get the milky white coloured fish soup, it is a must to add boiling water to the fish head while pan fry the fish and bring to vigorous boil with high heat.

Ingredients (Serves 2-4)

1 big fish head (about 1 catty, chopped to pieces)

1 head mustard green, insed and cut into pieces

8 straw mushrooms, halved

2 bricks firm tofu (cut into big pieces)

4 big slices ginger

1 tablespoon chopped spring onion

Seasonings

2 teaspoons ground pepper

1 teaspoon salt

1 teaspoon sugar

2 teaspoons rice wine

做法

1) 魚頭先洗淨抹乾。中火、鑊下油，先爆香薑片，再下魚頭煎至金黃，再圍邊灑少許米酒略炒至揮發。
2) 倒入約 800 毫升的熱滾水入鑊中，轉大火略滾 2 分鐘，令湯變奶白色。
3) 加入芥菜膽、豆腐及草菇，水滾後轉中火，蓋上蓋多煮 12 分鐘。
4) 最後加入胡椒粉、鹽及糖，調至自己喜歡味道便可熄火，再灑上葱花，即成。

Steps

1) Rinse and pat dry fish head. Heat the wok over medium heat. Add some oil. Stir-fry ginger until fragrant. Pan-fry fish head until golden. Drizzle with a little rice wine around the edge of wok. Stir-fry until wine has evaporated.
2) Add 800ml boiling hot water. Adjust to high heat and simmer for 2 minutes until soup turns milky white.
3) Add mustard green, tofu and straw mushrooms. Bring to boil. Adjust to medium heat. Cover the wok and cook for 12 minutes.
4) Add ground pepper, sugar and salt to get the desired taste. Turn heat off. Sprinkle with spring onion. Serve.

剁椒魚頭

Chopped Chilli Fish Head

四川名菜之一，
看上去像很辣，但其實並不是。
賣相也鮮豔吸引。

This is one of the famous Sichuan dishes.
Though it looks spicy, but it is not.
The presentation is colorful and appetizing.

材料 (2-4 人份量)

大魚頭 2 個 (開邊)
剁椒醬 1 飯碗
蒜頭 1 大個 (切碎)
薑 1 湯匙 (切碎)
葱花 1 棵 (切碎)
指天椒 2 隻
(喜歡吃辣的可加，不吃得太辣不用加)

調味料

鹽 半茶匙
生粉 1 茶匙
油少許
豉油少許

鹽不要下太多，因剁椒本身也很鹹。
Very little salt is required as chopped chilli sauce is already very salty.

Ingredients (Serves 2-4)

2 big fish heads, halved
1 rice bowl chopped chilli sauce
1 bulb garlic, finely chopped
1 tablespoon ginger, finely chopped
1 stalk spring onion, chopped
2 fresh red chillies
(optional for those who are not fond of spicy food)

Seasonings

1/2 teaspoon salt
1 teaspoon potato starch
A little oil
A little soy sauce

剁椒醬也可自己製造，方法轉載到第235頁的「Bonnie入廚小貼士」內。

Feel free to make your own chopped chilli sauce. Refer to "Bonnie's Kitchen Tips" on p.235.

做法

1) 魚頭洗淨，加入鹽及生粉醃一會。
2) 下少許油落鑊，爆香蒜頭，薑，加入剁椒醬和指天椒炒香炒勻。
3) 把已炒香的剁椒醬鋪在魚頭上，醃一會 (時間愈長愈入味)。
4) 隔水大火蒸 12 分鐘。
5) 鋪上葱花，淋熟油及少許豉油，即成。

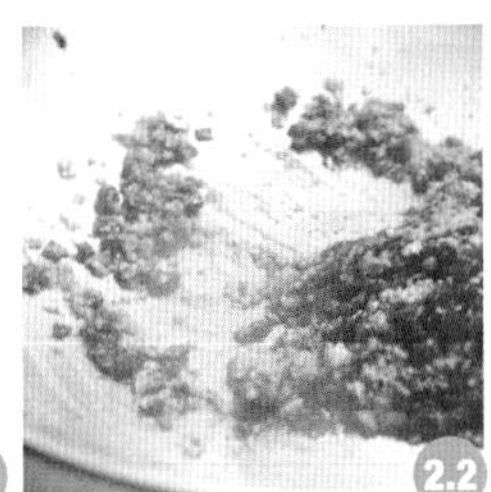

Steps

1) Rinse fish head. Marinate briefly with salt and potato starch.
2) Heat some oil in wok. Stir-fry garlic and ginger until fragrant. Add the chopped chili sauce and fresh red chilies. Stir-fry until fragrant.
3) Spread the chili mixture cover on the fish head. Let it marinate for a while (the longer the flavor is better absorbed)
4) Steam over with the high heat for 12 minutes.
5) Add chopped spring onion. Drizzle with cooked oil and a little soy sauce. Serve.

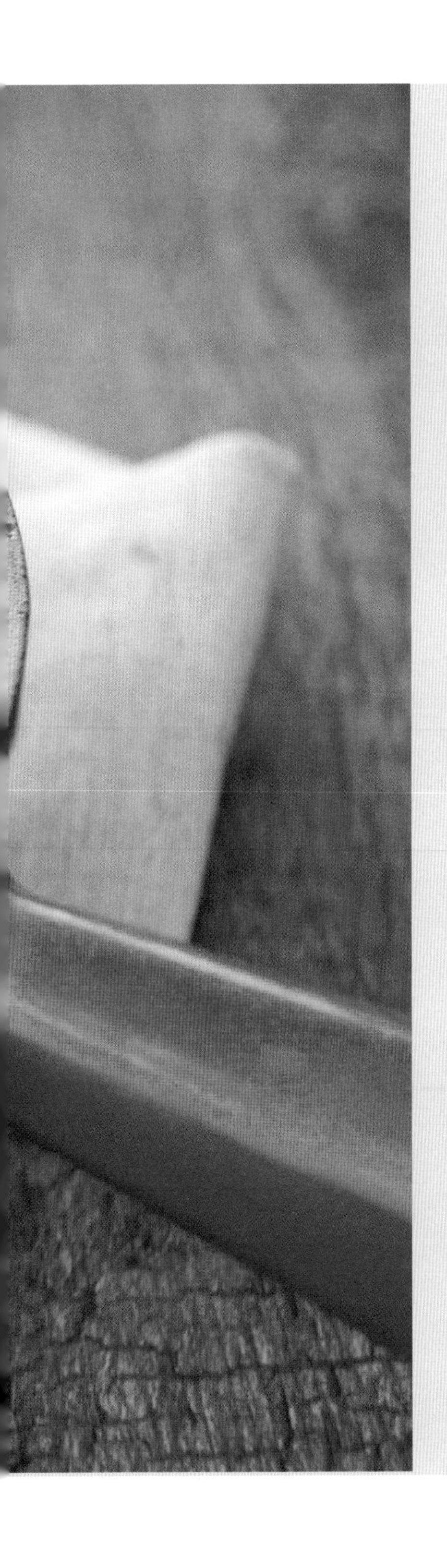

藥材醉蝦

Drunken Prawns with Chinese Herbs

這藥膳菜式，其實又是一個 10 分鐘便可煮好的餸，
材料易找，不花時間準備，
容易且極之好吃有益，秋冬天尤其適合。

This Chinese medicinal dish you can be cooked in 10 minutes. This dish has a lot of plus points: the ingredients are easy to find, requires very little time to prepare, delicious and healthy. It is most suitable for consumption during autumn and winter.

材料（2-4 人份量）

中蝦 12 隻

杞子 20 粒

當歸半塊（約半隻手掌般大小）

紅棗 6 粒（切半去核）

黃芪 2 條（切短）

雞湯 500 毫升

花雕 / 紹興酒 60 毫升

芫荽少許

糖 1 茶匙

煮蝦最理想的是九成熟，太熟會過韌。

Prawn tastes best when it is 90% cooked. When overcooked, it tastes rubbery.

Ingredients (Serves 2-4)

12 medium prawns

20 Chinese wolfberries

1/2 piece Dang Gui (size of half a palm)

6 red dates (halved and seeded)

2 sticks Huang Qi (sectioned)

500ml chicken broth

60ml Hua Diao / Shaoxing wine

A little coriander

1 teaspoon sugar

做法

1) 藥材用水沖洗一下，放進煲加入雞湯，用中火煮 3 分鐘至藥材出味。
2) 加入蝦，多煮約 1 分鐘七成熟。
3) 最後加入糖及花雕或紹興酒，多煮 1 分鐘至酒精略揮發便可。
4) 最後加些芫荽作裝飾，即成。

Steps

1) Rinse the herbs. Put into a pot. Add the chicken broth. Cook over in medium heat for 3 minutes until the herb flavors are released.
2) Add the prawns. Cook for 1 minute until medium well done.
3) Add the sugar and wine. Cook for 1 minute to evaporate off some of the alcohol.
4) Add coriander as decoration on top. Serve.

酒太早下酒味會盡失，太遲下又會因酒精重而苦澀。最後 1 分鐘下剛好。

Adding wine at the early stage will lose the wine aroma. Adding it too late will cause the dish to become bitter because of the alcohol. 1 minute before the dish is done is just right.

Chinese Delicacy
中式滋味

砵酒焗生蠔

Pan Fry Oysters with Port Wine

這道菜是我很喜歡的，
已經不是有太多地方可以吃到，
做出來甜甜的，又沒有酒味，
小朋友也可以吃。

This is one of my favourite dishes
but no longer served in most of restaurants.
It has a sweet flavour without the strong wine taste.
Even children can enjoy it.

材料（2-4 人份量）

生蠔 12 隻
洋葱半個（切條）
乾葱 1 個（切碎）
蒜頭 2 瓣（切碎）
牛油 30 克

Ingredients (Serves 2-4)

12 fresh oysters
1/2 onion (shredded)
1 clove shallot (chopped)
2 cloves garlic (chopped)
30g butter

調味料

砵酒 100 毫升
豉油 1 湯匙
蠔油 1 湯匙
糖 1 茶匙
生粉 6 湯匙
白胡椒粉 1 茶匙

Seasonings

100ml port wine
1 tablespoon soy sauce
1 tablespoon oyster sauce
1 teaspoon sugar
6 tablespoons potato starch
1 teaspoon white pepper powder

煮融牛油千萬不要太大火，否則很易燒焦。

Do not melt butter over high heat. It gets burned quickly.

做法

1) 生蠔洗淨，飛水 1 分鐘馬上撈起。用廚房紙抹乾，下少許胡椒粉，再沾上生粉。
2) 下牛油，用中大火煎香脆生蠔外表，取出備用。
3) 中小火下牛油，再下洋葱、乾葱、蒜碎，轉中火炒香。
4) 加入砵酒、糖、豉油及蠔油略煮至汁收杰。再把蠔回鍋炒勻，再蓋上蓋中小火焗煮 2 分鐘。食前放些芫荽裝飾即可。

2

4.1

4.2

Steps

1) Rinse oysters and blanch with boiling water for 1 minute. Drain immediately and pat dry with kitchen paper. Marinate with pinch of white pepper powder and coat the oysters with potato starch.
2) Pan-fry oysters with some butter over medium high heat until the surface is crispy fragrant. Set aside the oysters.
3) Heat some butter with medium low heat. Stir-fry the onion, shallot and garlic over medium heat until fragrant.
4) Add the port wine, sugar, soy sauce and oyster sauce. Cook until the sauce thickens. Add oysters. Stir until even. Cover wok and braise over medium low heat for 2 minutes. Add coriander just before serving.

Chinese Delicacy
中式滋味

白胡椒
香煎鮑魚

Pan Fry Abalone with White Pepper Sauce

我們很多時都愛用黑胡椒和白胡椒，
究竟應該怎樣配合食材呢？
其實白胡椒偏向跟海鮮比較夾，
黑胡椒則肉類為多，
而配上這鮑魚可說是最絕配。

Black and white peppers are common seasoning.
Which pepper should go with what ingredients?
Black pepper is more for meat whereas
white is more for seafood,
making it the perfect match with abalone.

材料（2-4 人份量）

鮮鮑魚 8 隻（去殼）

薑 5 片

葱 1 棵（切段）

月桂葉 1 片

雞湯 400 毫升

牛油 10 克（約 1 湯匙）

調味料

白胡椒粒 2 湯匙（壓碎）

糖 1 茶匙

生粉少許

Ingredients (Serves 2-4)

8 fresh abalones (shelled)

5 slices ginger

1stalk spring onion (sectioned)

1 bay leaf

400ml chicken broth

10g butter (about 1 tablespoon)

Seasonings

2 tablespoons white peppercorn (crushed)

1 teaspoon sugar

Pinch of potato starch

如用大連鮑燜煮 5 分鐘便可、南非鮑則需要更長時間煮至腍身，但只需 8 分鐘便會煮熟，可視乎個人喜好去決定吃爽身或腍身而調節烹調時間，最重要燜煮期間一定要用最小火，否則鮑魚會韌。

If Dalian abalones are used, cook for 5 minutes will do. South African abalones will take longer to become tender but only 8 minutes to get cooked. Feel free to adjust the cooking time according to personal preference for either a crunchier or tender mouth-feel. It is most important to braise the abalones with the lowest heat so that the meat texture will not become too tough.

做法

1) 鮑魚去殼洗刷乾淨，用煲把雞湯煮滾，加入 1 湯匙白胡椒及月桂葉，再放入鮑魚轉最小火煮 5-15 分鐘至鮑魚熟及腍身。
2) 煮好的鮑魚拿起抹乾備用，雞湯隔去渣留起備用。
3) 中小火下牛油，溶化後轉大火放入鮑魚，快速煎香兩面至金黃便可，取出備用。
4) 中火下油，放入白胡椒碎，慢炒約 2 分鐘至有胡椒氣味散出後，轉中大火加入薑葱炒香。
5) 加入約 150 毫升之前煮過鮑魚的雞湯，再加入糖，煮滾後再加入鮑魚炒勻，最後加入生粉水打芡至汁醬掛身即成。

煎鮑魚時每面煎的時間不要超過 30 秒，否則鮑魚會過老。

When pan-frying the abalones, do not exceed 30 seconds per side to prevent abalone texture from becoming too tough.

Steps

1) Remove the abalone shell and scrub the abalones. Bring the chicken broth to boil in a pot. Add 1 tbsp white peppercorn and the bay leaf. Add abalones. Cook over the lowest heat for 5-15 minutes until abalones are tender and cooked.
2) Dish up the abalones, pat dry. Strain and sieve the chicken broth and set aside.
3) Heat the wok with medium low heat. Add butter. Add abalones when butter has melted. Pan-fry swiftly until both sides of abalones become golden. Set aside.
4) Heat the wok with medium heat. Add some oil. Stir-fry crushed white peppercorn slowly for 2 minutes until fragrant. Adjust to medium high heat. Add ginger and spring onion. Stir-fry until fragrant.
5) Add 150ml chicken broth previously used for cooking abalones. Add sugar. Bring to boil. Add abalones. Stir-fry until even. Add some potato starch solution. Stir well until abalones are coated with sauce.

1

3

4

5

過橋象拔蚌

Poached Geoduck Clam

我自己最喜歡吃海鮮，象拔蚌更是最愛之一。
其實它並不難處理，而且在家吃價錢便宜得多，
這裏讓我示範一下如何劏象拔蚌
和煮出這道名貴的菜式吧！

Seafood is my favourite and geoduck clam tops the list.
It is not too difficult to handle and cooking it personally
will greatly reduce the cost.
Let me show you how to dress geoduck clam
and the proper way of cooking this expensive dish.

材料（2-4 人份量）

象拔蚌約 1 斤半
蝦頭、蝦殼、蚌膽約 300 克
（約有 14 隻蝦頭加蝦殼）
甘筍 2 條（切粒）
西芹 1 條（切粒）
銀芽 150 克
蒜頭 3 瓣（拍扁）
乾葱 4 粒（切半）
白胡椒粉 2 茶匙
芫荽少許

調味料

茄膏 2.5 湯匙
糖 1 茶匙
鹽 1 茶匙
紹興酒 1 湯匙

「過橋」亦即是蚌片不用煮，滾湯過一下便可，吃時因切得薄不會太生，甚至熟。

Lightly poach geoduck will do. Slice it thinly so that the meat is not too raw or likely cooked.

Ingredients (Serves 2-4)

1.5 catties geoduck clam

300g prawn head, prawn shell and oval stomach of geoduck clam

(about 14 prawn heads and shell)

2 carrots (diced)

1 stick celery (diced)

150g bean sprout

3 cloves garlic (smashed)

4 cloves shallot (halved)

2 teaspoons white pepper powder

A little coriander

Seasonings

2.5 tablespoons tomato paste

1 teaspoon sugar

1 teaspoon salt

1 tablespoon Shaoxing wine

湯底是整個餸的靈魂，一定要煮得夠鮮，味道可濃一點點，因配銀芽後味道會減淡一點。

Seafood broth is the soul of this dish. It must be made fresh. The flavor can be on the heavy side as the taste will be lightened by the bean sprout.

食時可加入切片的隔夜油條，吃時會更加香口。

Add some sliced overnight fried dough stick Youtiao upon serving to enhance the taste.

做法

1) 象拔蚌先去殼，再浸入熱滾水 2 分鐘，便可撕去衣，用食水再沖洗一下，找熟食切板，放横平中間切開，打開放平，再平斜刀橫切薄片，到最尾為止。
2) 切好的象拔蚌片放入雪櫃待用。
3) 中大火下油，先炒蝦殼、蝦頭及蚌膽，炒至金黃色，香味散出。之後圍邊加紹興酒。再炒香蒜及乾葱。最後放入甘筍及西芹略炒。
4) 加入 1 公升水、胡椒粉。大火開蓋煮 20 分鐘至湯收至約剩 600 毫升。最後下所有調味料，試味。隔渣，留湯，備用。
5) 銀芽另外用滾水煮熟，放在大碟底部，平鋪好。
6) 大火把海鮮湯底翻熱煮滾，湯滾後加入蚌片，隨即馬上熄火。把湯連蚌片倒在芽菜上面，再灑些芫荽葱上面，即成。

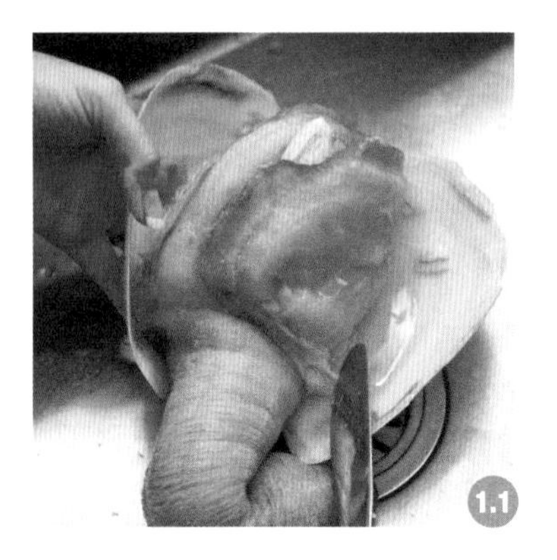
1.1

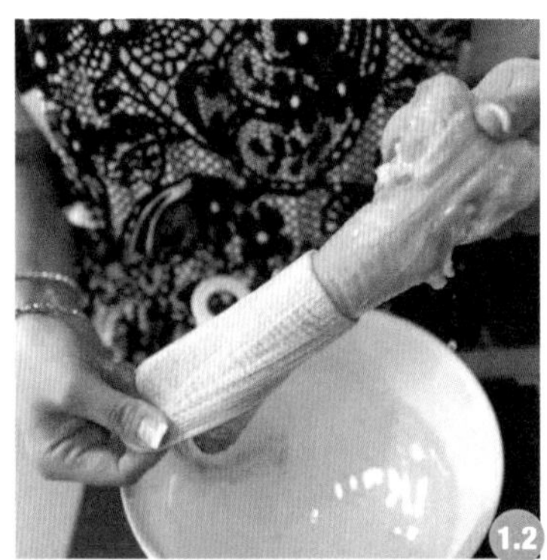
1.2

1.3

Steps

1) Remove the shell of the geoduck. Soak it into boiling water for 2 minutes. Peel off the membrane and rinse it with some drinkable water. Place the geoduck on a chopping board for cooked food. Cut into half lengthwise. Thinly slice it aslant to the end.

2) Chill the sliced geoduck in the fridge.

3) Heat oil in the wok over medium high heat. Stir-fry the prawn head, prawn shell and oval stomach of geoduck until golden and fragrant. Add Shaoxing wine. Stir-fry garlic and shallot until fragrant. Add carrot and celery. Stir-fry briefly.

4) Add 1 liter of water and the ground pepper. Cook over high heat for 20 minutes (do not cover the wok) until stock is reduced to around 600ml. Add seasoning ingredients. Taste the broth. Strain ingredients with sieve and set aside the broth.

5) Blanch bean spout in boiling water until done. Place in a big plate.

6) Bring seafood broth to boil with high heat. Add sliced geoduck. Turn the heat off immediately. Pour the geoduck with broth on the bean sprout. Sprinkle with some coriander and spring onion. Serve.

這海鮮湯底也可以做象拔蚌、龍蝦或海鮮泡飯，海鮮加些蔬菜粒再淋上飯上便成。

This seafood broth can also be the soup base for rice with geoduck, lobster or seafood. Add some diced of vegetable and seafood. Drizzle over rice.

若沒有蝦殼，可用原隻小蝦連頭連殼去做。

If there is no prawn shell, cook the broth with whole tiny shrimps with shell and head instead.

Chinese Delicacy
中式滋味

水煮魚

Boiled Fish in Sichuan Style

這個四川名菜，賣相看上去十分霸氣，
紅紅的辣椒，大大一盆的，
但你有沒有想過其實在家做十分簡單，
材料也不太複雜。

This famous Sichuan dish with its big pot of red chilies
does look overwhelming.
The truth is it is not that difficult to make it at home
and the ingredients are not too complicated.

材料（2-4 人份量）

龍脷柳 500 克（切塊）
豬紅 1 磚（切條再浸水）
大豆芽 300 克（洗淨浸水）
乾辣椒約 30 隻
指天椒 3 隻
花椒 3 茶匙
薑片 5 大片
蒜頭 6 瓣（拍扁）

調味料

桂林辣椒醬 1.5 湯匙
辣豆瓣醬 1.5 湯匙
蠔油 1 湯匙
糖 2 茶匙
花椒油 3 湯匙
雞湯 600 毫升

龍脷柳醃料

油 3 湯匙
糖 1 茶匙
鹽 1 茶匙
紹興酒 1 湯匙
蛋白 1 隻（打勻）

這是小辣版本，喜歡更辣的可多加指天椒，更麻便多加花椒油。喜歡比較油的你也可以自己多加些油去炸花椒。

This is a mildly spicy version. For those who like it spicy, feel free to add more fresh red chilies; add more Sichuan pepper oil for those like the "lull" flavor and add more oil to deep-fry the Sichuan peppercorn for those who like it oilier.

Ingredients (Serves 2-4)

500g sole fillet (cut into pieces)

1 brick pig blood curd (cut into strips, soak in water)

300g soy sprout (rinse, soak in water)

30 dried chillies

3 fresh red chillies

3 teaspoons Sichuan peppercorn

5 big slices ginger

6 cloves garlic (smashed)

Marinade for sole fillet

3 tablespoons oil

1 teaspoon sugar

1 teaspoon salt

1 tablespoons Shaoxing wine

1 egg white (beaten)

配菜除了豬紅外，你亦可以放山筋、腐皮豆卜、薯粉、蒿荀等……薯粉要先浸水才煮，以上配料大約煮 4-8 分鐘便可。

Aside from pig blood crud, feel free to add some deep-fried gluten balls, tofu skin, tofu puff, tapioca noodle or Chinese cabbage in it. Soak the tapioca noodle in the cold water for a while before adding to the cooking. Cook these side ingredients for 4-8 minutes can be done.

Seasonings

1.5 tablespoons Guilin chili sauce

1.5 tablespoons spicy chili bean sauce

1 tablespoon oyster sauce

2 teaspoons sugar

3 tablespoons Sichuan peppercorn oil

600ml chicken broth

做法

1) 龍脷柳放入所有醃料醃一會，備用。
2) 鑊下油，中大火先炒香蒜頭、薑及指天椒。再加入桂林辣椒醬及豆瓣醬炒香。
3) 放入雞湯煮滾。加入蠔油、花椒油及糖調味。
4) 把豬紅或你喜愛的配菜加入，繼續煮大約 5 分鐘至豬紅和配料熟及入味。
5) 另外一個鍋煮起滾水，把大豆芽煮熟，約需 4 分鐘。煮好的大豆芽鋪在另一鍋底，備用。
6) 另一個小鍋開小火，凍鑊下 6 湯匙油及乾辣椒，略炸 3 分鐘，再加入花椒至炸香味散出便可。備用。
7) 豬紅及配料煮至熟及入味後，加入龍脷柳略煮 2 分鐘至剛熟便熄火。
8) 把整鍋倒入放了大豆芽的鍋中。
9) 最後加上剛炸好的辣椒及花椒連油，再在上面放些芫荽葱即成。

Steps

1) Marinate the sole fillet with all the marinade ingredients. Set aside.

2) Heat oil in the wok. Stir-fry garlic, ginger and fresh red chilies over medium high heat until fragrant. Add Guilin chili sauce and chili bean sauce. Stir-fry until fragrant.

3) Add the chicken broth. Bring it to boil. Add oyster sauce, Sichuan pepper oil and sugar as seasoning.

4) Add the pig blood curd or other preferred side ingredients. Cook for 5 minutes until ingredients are cooked and have absorbed the flavors.

5) Boil some water in another pot. Blanch soy sprouts until cooked, takes about 4 minutes. Transfer to a casserole. Set aside.

6) Add 6 tablespoons oil and dried chilies to another a small pan. Warm over low heat. Deep-fry the dried chilies for 3 minutes. Add the Sichuan peppercorn and deep-fry until fragrant. Remain it hot with low heat.

7) Add the sole fillet at last and cook for 2 minutes until just cooked. Turn the heat off.

8) Transfer all the content in the whole the wok to the casserole containing soy sprout.

9) Add the fried chilies and Sichuan peppercorn with the oil all on top on the fish, put some coriander. Serve.

不煮魚可改用牛或田雞。若煮牛在醃肉時請勿下鹽。鹽會令牛肉變韌。

Fish fillet can be replaced by beef or frog. If using beef, do not add salt to the marinating as salt will make the beef texture become chewing.

花椒十分容易炸焦，要最後下及全程小火。

As Sichuan peppercorn easily gets burned, the entire deep-frying process should be done over low heat and the peppercorn is to be added last.

Chinese Delicacy
中式滋味

京葱燒雞

Braised Chicken with Leek

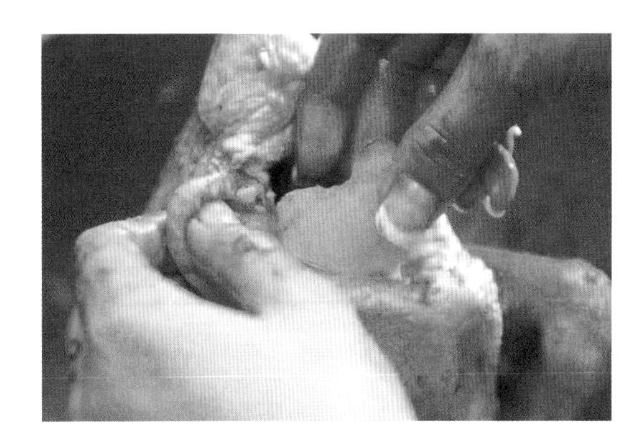

跟師傅鼎爺學的一道菜式，
沒有做「乞兒雞」那樣繁複，卻美味不減。
自己再改良簡化一點，
做成這個美味易做的宴客佳餚。

This is a dish that I learned from my master Steve Lee.
It is not as complicated as Beggar's Chicken
but the taste is equally delicious. I have simplified
the process slightly to create this guest worthy dish.

材料（4人份量）

雞1隻（洗淨抹乾）
京葱2棵（切斜段）
葱3棵（2棵打結，1棵切段）
薑6大片（其中拿1片切薑蓉）
蒜頭6瓣（4粒拍扁，2粒切蓉）
乾葱1粒（切蓉）
玫瑰大頭菜2片（切碎）

醃料

粗鹽2茶匙
糖2茶匙
五香粉半茶匙
油2茶匙
老抽2茶匙

調味料

片糖半片（敲碎）
蠔油2湯匙
生抽2茶匙
紹興酒4湯匙
生粉少許

Ingredients (Serves 4)

1 chicken (rinsed and pat dry)

2 leeks (sectioned)

3 stalks spring onion

(tie 2 stalks into knots, cut 1 stalk into sectioned)

6 big slices ginger (minced 1 slice)

6 cloves garlic (smash 4 cloves, minced 2)

1 clove shallot (minced)

2 big pieces pickled mustard root/ Datoucai (finely chopped)

Marinade

2 teaspoons coarse salt

2 teaspoons sugar

1/2 teaspoon five spice powder

2 teaspoons oil

2 teaspoons dark soy sauce

Seasonings

1/2 piece brown slab sugar (smashed)

2 tablespoons oyster sauce

2 teaspoons soy sauce

4 tablespoons Shaoxing wine

Pinch of potato starch

最後階段再加入些綠色京葱令菜式更具色香味。

Adding some green sections of leek at the final stage to enhance the aroma and presentation.

煮雞要用小火，大火會令雞肉容易變老。

Must use the low heat to braise the chicken. High heat will result in tough meat texture.

做法

1) 薑蓉、乾葱蓉及蒜蓉加入所有醃料（除了老抽）拌均，塗在雞的內外。再把薑片、一半份量的京葱及打了結的葱放入雞腔內再用長針封起。最後放入玫瑰大頭菜及老抽擦均，醃最少 30 分鐘（時間愈長愈入味）。
2) 預熱焗爐到 220 度，放入雞焗 20 分鐘，燒至外皮金黃香脆後拿出。
3) 用深鍋大火下油，先炒香薑片、蒜頭及京葱，再下片糖略炒。加入 500 毫升的水及所有調味料攪均，再加入少許生粉水令湯汁濃稠一點，放入雞，汁再滾起後蓋上鍋蓋轉小火多煮 10 分鐘，再加入幾段綠色京葱後熄火讓焗雞多 10 分鐘至雞熟。
4) 把雞斬件後再淋上汁上碟即成。

若家中沒有焗爐，可改用走油或煎去令雞皮香脆。

If there is no oven at home, deep fry or pan fry the chicken briefly until the skin is crispy and golden.

焗雞時可中間時段反雞一下，令其燒得更平均。

Half way into the baking, turn chicken over to let the chicken bake more evenly.

Steps

1) Combine minced ginger, shallot and garlic with all the marinade ingredients (except dark soy sauce). Spread evenly all over the chicken including the cavity. Stuff ginger, half of the leek and the knotted spring onion into chicken cavity. Seal with a long needle. Rub the chicken with the pickled mustard root and dark soy sauce. Marinate it for at least 30 minutes (the longer it is the better to absorb the flavours).

2) Preheat the oven to 220°C. Put the chicken and bake for 20 minutes until the skin is golden and crispy.

3) Heat a deep pot with high heat. Add oil. Stir-fry the rest ginger, garlic, leek and spring onion until fragrant. Add brown slab sugar. Stir-fry briefly. Add 500ml water and seasoning ingredients. Stir well. Add a little potato starch solution to thicken the sauce. Add chicken. Bring to boil. Cover pot and simmer over low heat for 10 minutes. Add a few green sections of leek. Turn heat off and let it soak for 10 minutes until chicken is done.

4) Chop the chicken into pieces. Transfer to plate and drizzle with sauce. Serve.

辣子雞

Spicy Chilli Chicken

辣子雞又香又辣，是一道很好下飯的菜式。
它的顏色吸引，絕對是色、香、味俱全。
自己煮亦可以控制辣的程度，
乾辣椒看似很辣但其實香味居多，
吃得辣的加多點指天椒。

This is an appetizing dish that is spicy and aromatic,
with a attractive colour.
The level of spiciness can be controlled when it is home-cooked.
The dried chillies may look really spicy
but it is used more for its unique aroma.
For those who like it spicy, add more red chillies.

材料（2-4 人份量）

鮮雞半隻
（斬細件）
乾辣椒 50 克
指天椒 1 隻（切粒）
乾葱 2 粒（切碎）
蒜頭 4 瓣（切碎）

Ingredients (Serves 2-4)

1/2 fresh chicken
(chopped to small pieces)
50g dried chillies
1 fresh chilli (diced)
2 cloves shallot (finely chopped)
4 cloves garlic (finely chopped)

調味料

椒鹽 2 茶匙
孜然粉半茶匙
花椒油 1 湯匙
鹽 1 茶匙
糖 1 茶匙
生粉 4 湯匙

Seasonings

2 teaspoons pepper salt
1/2 teaspoon cumin powder
1 tablespoon Sichuan pepper oil
1 teaspoon salt
1 teaspoon sugar
4 tablespoons potato starch

醃料

豉油 1 湯匙
鹽 1 茶匙
糖 1 茶匙
麻油 1 湯匙
紹興酒半湯匙
粟粉 1 茶匙

Marinade

1 tablespoon soy sauce
1 teaspoon salt
1 teaspoon sugar
1 tablespoon sesame oil
1/2 tablespoon Shaoxing wine
1 teaspoon corn starch

想辣一點可以加多 1、2 隻指天椒，不想太辣亦可以不下。

To increase the level of spiciness, add 1 or 2 more fresh red chillies. It can be omitted if you preferred to keep the dish less spicy.

可用田雞或蝦取代雞，若煮蝦要去殼開邊，醃料可除去豉油。

Chicken can be replaced by the frog or prawns. When dealing with the prawns, remove the shell and slit back of prawns, remove soy sauce from the marinade as well.

做法

1) 雞件洗淨抹乾，放入所有醃料醃最少 30 分鐘（愈久愈入味）。
2) 醃好的雞件撲上薄薄的生粉。
3) 中火下多點油把雞件走油，油溫至 180℃便可以加入雞件炸至金黃色，雞件瀝油備用。
4) 中大火下油，先下乾葱、蒜蓉及指天椒炒香，再加入乾辣椒炒香。
5) 把雞件回鑊略炒，加入調味料（花椒油除外）炒熟，最後加入花椒油炒均，即成。

Steps

1) Rinse and wipe dry the chicken pieces. Combine with all the marinade ingredients and let them stand for minimum 30 minutes. (the longer the better to absorb the flavor)
2) Dust the marinated chicken pieces with a thin layer of potato starch.
3) Heat enough oil over medium heat to 180°C. Put in the chicken pieces and deep-fry to golden colour. Dish up the chicken pieces. Set aside oil.
4) Heat little oil over medium high heat. Stir-fry the shallot, garlic and fresh red chillies until fragrant. Add the dried chillies. Stir-fry until fragrant.
5) Add the chicken pieces. Stir-fry briefly. Add the seasoning ingredients (except the Sichuan pepper oil). Stir-fry until done. Add the Sichuan pepper oil. Stir until even. Serve.

麻辣雞煲

Sichuan Spicy Chicken Casserole

很多人也喜歡吃雞煲，其實煮法十分簡單。
食法也有兩種，下多點酒及雞湯
便可以做火鍋湯底，
要乾身一點下少一點湯汁便可，任君選擇。

This dish is the favourite of many people.
The cooking steps are rather simple.
It can be served two ways according to your personal preference.
Adding more wine and chicken broth will turn into
a hotpot soup base whereas reducing the broth
will make it into a dish with thick sauce.

材料（2-4 人份量）

鮮雞 1 隻（斬件）
乾葱 4 粒（一開二）
蒜頭 5 瓣（拍扁）
薑片 5 片
指天椒 2 隻（切粒）
乾辣椒 10 隻
芫荽（少許）
葱段（少許）

調味料

麻辣醬 2 湯匙
花椒油 2 湯匙
紹興酒 250 毫升
老抽半湯匙（調色用）
糖 1 茶匙
豉油少許（不夠鹹才加）
雞湯 100 毫升

雞要用小火燜煮才不易變老。

Braise the chicken with low heat so that the meat will not become tough.

醃料

豉油 2 湯匙
老抽 1 湯匙
紹興酒 1 湯匙
糖 2 茶匙
麻油 1 湯匙

Ingredients (Serves 2-4)

1 fresh chicken (chop into pieces)

4 clove shallot (halved)

5 cloves garlic (smashed)

5 slices ginger

2 fresh red chillies (diced)

10 dried chillies

A little coriander

A little spring onion

Marinade

2 tablespoons soy sauce

1 tablespoon dark soy sauce

1 tablespoon Shaoxing wine

2 teaspoons sugar

1 tablespoon sesame oil

Seasonings

2 tablespoons Sichuan spicy sauce

2 tablespoons Sichuan peppercorn oil

250ml Shaoxing wine

1/2 tablespoon dark soy sauce

(for colouring)

1 teaspoon sugar

A little soy sauce

(to be added if the dish is not salty)

100ml chicken broth

怕酒味太重的可以用些雞湯去取代紹興酒。

For those who prefer a milder wine flavor, replace some of the Shaoxing wine with chicken broth.

做法

1) 雞件洗淨抹乾，放下所有醃料把雞醃最少 30 分鐘。
2) 熱鍋中火下少許油，先爆香乾葱、薑及蒜。加入麻辣醬、指天椒及乾辣椒炒勻。
3) 下雞件，轉大火。要把雞件外面煎香。不要炒來炒去，煎香一面再反另一面。
4) 加入紹興酒，雞湯炒勻；再放入花椒油。
5) 蓋上蓋轉小火煮 5 分鐘，再焗 10 分鐘。
6) 最後加上芫荽和葱段，即成。

2

Steps

1) Rinse and wipe dry the chicken pieces. Combine with all the marinade ingredients and let them stand for minimum 30 minutes.
2) Heat the pot over medium heat. Add some oil. Stir-fry the shallot, ginger and garlic until fragrant. Add the Sichuan spicy sauce, fresh red chillies and dried chillies.
3) Add the chicken pieces. Adjust to high heat. Pan-fry chicken pieces until the exterior is fragrant. Do not stir-fry. Pan-fry one side until fragrant before turning the chicken pieces over.
4) Add the Shaoxing wine and chicken broth. Stir well. Add the Sichuan peppercorn oil.
5) Cover the pot. Adjust to low heat and cook for 5 minutes. Turn heat off and let them stand for 10 minutes.
6) Add the coriander and spring onion. Serve.

3

5

Chinese Delicacy
中式滋味

鐵板黑椒鹹豬手

Sizzling Black Pepper Sauce with Salted Pork Knuckle

一個有名的大牌檔菜式，在家做也很輕鬆，
順便可學懂煮黑椒汁，以後用來配搭甚麼材料也沒有難度啦！

This famous street-restaurant dish can be easily prepared at home.
On top of that, you also get to learn to make black pepper sauce,
a great accompaniment to many other ingredients.

材料（2-4 人份量）

急凍鹹豬手 1 隻
洋葱半個（切條）
大紅椒 1 隻（切片）
蒜頭半個（切碎）
黑胡椒粒 6 湯匙（壓碎）
月桂葉 4 片

Ingredients (Serves 2-4)

1 frozen salted pork knuckle
1/2 onion (shredded)
1 red bell pepper (shredded)
1/2 bulb garlic (chopped)
6 tablespoons black peppercorn (crushed)
4 bay leaves

調味料

白醋 1 茶匙
雞湯 300 毫升
老抽 1 湯匙
蠔油 1 湯匙
糖 3 茶匙

Seasonings

1 teaspoon white vinegar
300ml chicken broth
1 tablespoon dark soy sauce
1 tablespoon oyster sauce
3 teaspoons sugar

若沒有鐵板，可放入鑄鐵鍋中或直接上碟。

If the hot plate is not available, place the pork knuckle in cast-iron pot or on a serving dish.

做法

1) 豬手解凍洗淨，放入蓋過面的滾水內，加入2湯匙黑椒碎、月桂葉和白醋。水滾後轉小火蓋上蓋燜 2.5 小時。之後熄火讓豬手焗最少 1 小時或以上至食前才拿出來。
2) 把豬手放在鐵板上，用小火燒熱。
3) 另一面便做黑椒汁。中火下油，加入 4 湯匙黑椒碎略炒至香味散出，再加入蒜蓉、大紅椒及洋葱炒香。
4) 放入雞湯、老抽、蠔油、糖。略煮一會至洋葱軟身及汁帶少許濃。加入生粉水打芡。最後把汁淋在鐵板上的豬手即成。

Steps

1) Defrost, wash then wipe dry the pork knuckle, add boiling water to cover the pork knuckle. Add 2 tablespoons crushed black pepper, bay leaf and white vinegar. Bring to boil. Adjust to low heat. Cover the pot. Braise for 2.5 hours. Turn heat off. Let them soak for 1 hour or more. Remove the pork knuckle before serving.

2) Place the pork knuckle on a hot plate. Heat it with low heat.

3) Making the black pepper sauce: heat the pan over medium heat, add some oil. Add 4 tablespoons crushed black pepper. Stir-fry briefly until fragrant. Add the garlic, red bell pepper and onion. Stir-fry until fragrant.

4) Add the chicken broth, dark soy sauce, oyster sauce and sugar. Cook for a while until the onion becomes tender and the sauce thickens. Stir in a mixed potato starch with water. Drizzle sauce over the hot plate. Serve.

酸辣土豆絲

Sour and Spicy Shredded Potato

第一次在國內吃到這道菜時，真是估不到原來用馬鈴薯也可以做得這樣爽口，做法又十分之簡單，更可用少於十元的價錢做出一碟美味菜式，很讚。

The first time I ate this dish in China, I could not believe that potato could taste as crunchy as this. The process is really simple and the cost is less than ten dollars. Awesome!

只下鎮江醋不夠酸，所以要加白醋增加酸味。

Zhenjiang vinegar alone is not that sour. White vinegar is added to enhance the flavor.

材料 (2-4 人份量)

馬鈴薯 1 個 (去皮後刨或切幼條)
乾辣椒 3、4 隻 （切粒）
指天椒 1 隻
(切粒，如不喜歡太辣可以不下)

Ingredients (Serves 2-4)

1 potato (peeled, grated or cut into thin shreds)
3 or 4 dried chillies (diced)
1 fresh red chilli
(diced, optional for those who prefer it to be less spicy)

調味料

鎮江醋半湯匙
白醋 1 茶匙
豉油 1 湯匙
(可加可減，亦可用鹽取代)
花椒油 1 湯匙

Seasonings

1/2 tablespoon Zhenjiang vinegar
1 teaspoon white vinegar
1 tablespoon soy sauce (feel free to adjust the quantity or replace it with salt)
1 tablespoon Sichuan peppercorn oil

做法

1) 馬鈴薯絲用水過幾次去除澱粉質，再加清水浸最少半小時以上，瀝乾備用。
2) 中大火下油爆香辣椒，再加入馬鈴薯絲略炒 1 分鐘。
3) 下白醋，鎮江醋及豉油炒勻及炒熟，最後下花椒油炒勻，即成。

另外一做法是可以用白醋跟鹽取代鎮江醋及豉油，但用鎮江醋加豉油會香口些。

Another option is to replace Zhenjiang vinegar and soy sauce with the white vinegar and salt, though the former tastes better.

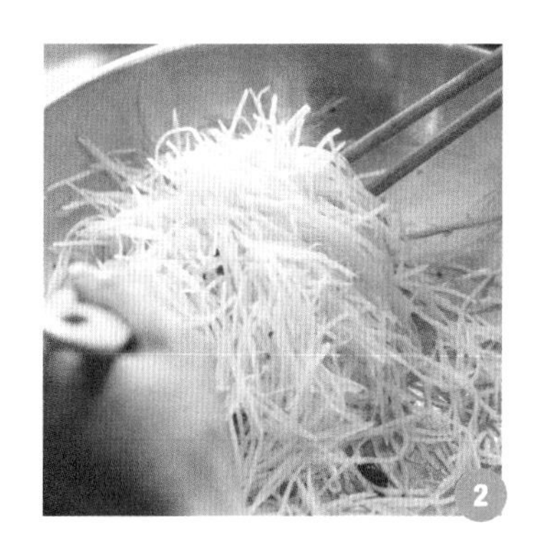

Steps

1) Rinse the shredded potato a few times to remove the starch. Soak it in the water for at least half an hour. Drain and set aside.
2) Heat the oil with medium high heat. Stir-fry the chillies until fragrant. Add the shredded potato and stir-fry briefly.
3) Add the white vinegar, Zhenjiang vinegar and soy sauce. Stir-fry until even and done. Add the Sichuan pepper oil. Stir well. Serve.

浸水程序就是令馬鈴薯爽口的秘訣。

The secret tip to that crunchy texture of shredded potato is the soaking process.

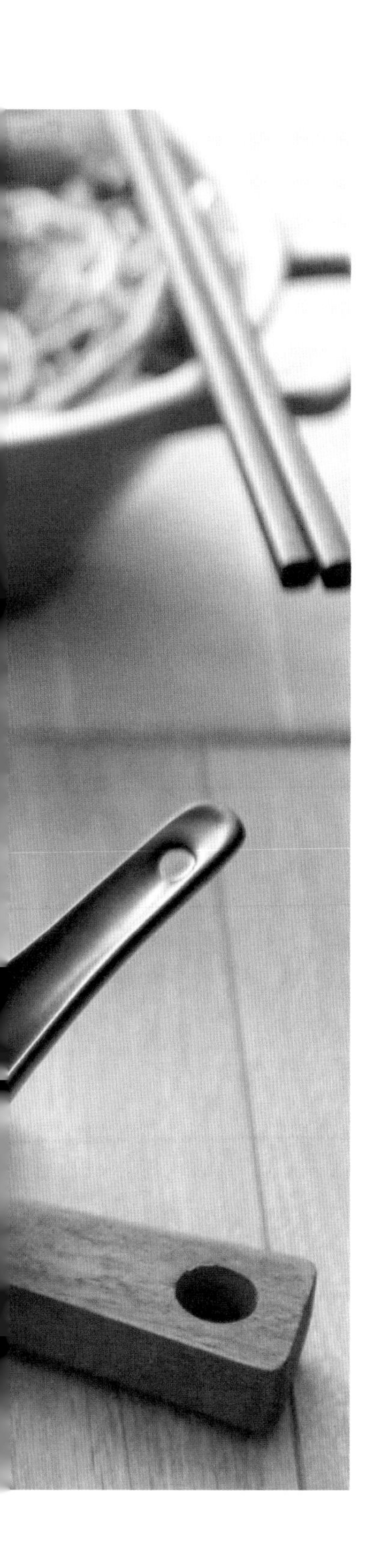

Chinese Delicacy
中式滋味

秘製
紅燒牛肉麵
Secret Recipe of Taiwanese Braised Beef Noodles

聽好友說她吃過最好吃的燜牛腩，便是她老爺曾經做過的。一次機緣在她家作客見到她的老爺，便大膽地向他請教。他十分慷慨地馬上找了那食譜給我，一看才知原來是個「紅燒牛肉」的食譜。我自己改良再加自製拉麵，不求人便做出比外面更真材實料無味精的無敵「紅燒牛肉麵」版本。再次多謝洪金寶大哥的無私分享！

A good friend told me that the best braised beef she has ever tasted was made by her father-in-law. When I had the opportunity to meet him at my friend's home. I plucked up the courage and asked him about it. He immediately searched for the recipe and shared it with me. It is actually the recipe for Taiwanese braised beef. After tweaking the recipe and adding my home-made ramen, it now becomes a Taiwanese braised beef noodle tastier than any that is served in the restaurants! A big Thank You to Uncle Sammo Hung for this sharing.

材料（4 人份量）

牛腩 / 牛腰 1.5 斤
牛骨半斤
白蘿蔔 1 條（切塊）
甘筍 1 條（切塊）
洋葱 1 個（切塊）
京葱 1 棵（切段）
橙 1 個（去皮）
蒜頭 1 個（拍扁）
老薑 5 大片
芫荽及葱段（少許）

香料

陳皮 1 片（浸軟洗淨）
草果 2 粒（拍碎，只要外殼）
肉寇 3 粒（拍碎）
八角 4 粒
花椒 2 茶匙
白胡椒粒 2 茶匙
甘草 3、4 片
小茴香 1 茶匙

調味料

四川郫縣辣豆瓣醬 3 大湯匙
豆瓣醬 1 大湯匙
豉油 30 毫升
老抽 10 毫升
冰糖 5 茶匙（約 1 粒）
米酒 6 湯匙
鹽 1 茶匙
糖 1 茶匙

拉麵材料

高筋麵粉 300 克
水 140 毫升
鹽 1/3 茶匙
油半茶匙

Ingredients (Serves 4)

1.5 catties beef brisket / beef shank

1/2 catty beef bone

1 radish (cut into pieces)

1 carrot (cut into pieces)

1 onion (cut into pieces)

1 stalk leek (cut into sections)

1 orange (peeled)

1 bulb garlic (smashed)

5 big slices old ginger

Coriander and spring onion (optional)

Spices

1 piece dried tangerine peel (soaked until tender and rinsed)

2 pieces Cao Guo (smashed)

3 pieces Nutmeg (smashed)

4 pieces Star Anise

2 teaspoons Sichuan peppercorns

2 teaspoons white peppercorns

3 or 4 slices licorice

1 teaspoon fennel seeds

Seasonings

3 big tablespoons Sichuan spicy chilli bean sauce

1 big tablespoon chilli bean sauce

30ml soy sauce

10ml dark soy sauce

5 teaspoons rock sugar (1 piece)

6 tablespoons Chinese rice wine

1 teaspoon salt

1 teaspoon sugar

Ingredients for Ramen

300g bread flour

140ml water

1/3 teaspoon salt

1/2 teaspoon oil

若不放入豆瓣醬、老抽及豉油，便會變成「清燉牛肉麵」。請加多點鹽去調味。

To make a braised beef noodle with clear soup style, do not add the chilli bean sauce, dark soy sauce and soy sauce. Instead, add a little more salt as seasoning to make the soup base more clear.

紅燒牛肉做法

1) 香料用白鑊中火烘幾分鐘至香味散出，備用。
2) 牛腩 / 牛�及牛骨出水，備用。中火下油炒香蒜、薑，再下兩種豆瓣醬炒香。
3) 加入牛腩 / 牛腱及牛骨炒勻，再圍邊繞入米酒略炒。之後放入豉油、老抽炒香。再加入白蘿蔔及甘筍炒勻。
4) 加入 2.5 公升水，把所有香料、京葱、橙、洋葱放入煲魚袋後再放入。加入冰糖，轉慢火燜 2 小時，最後如有需要可以下些鹽糖調味。
5) 把煮熟的麵放在碗中，加入熱湯、鋪上牛腩 / 牛腱及蘿蔔，再灑上芫荽葱，即成。

拉麵做法

1) 麵粉加入鹽及油，再慢慢加入水搓勻麵糰至光滑，用保鮮紙蓋着 1 小時去醒麵糰。
2) 多加點手粉在枱上，把麵糰搓扁，兩面撲少許粉，再放入切麵機切麵（若沒有切麵機可自已用刀切或自己拉）。
3) 把切好的麵放入滾水內煮約 2、3 分鐘至熟，拿起用食水過一下冷河，即成。

貼士
tips #2

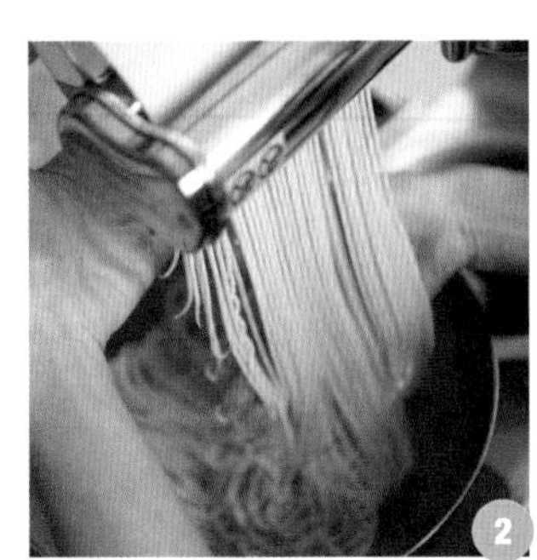

若想直接煮「紅燒牛肉」的話，請減少水的份量至 2 公升。洋葱、橙、京葱亦不用放入煲魚袋。亦不用下牛骨。這樣做湯汁會比較濃厚，配料入煲魚袋會比較清。

To make a braised beef dish without soup and noodle, reduce the water to 2 liters. And the onion, orange and leeks no need to put into the food filter bag. Beef bone can be removed. The sauce will become thicker. Keeping the side ingredients in the food filter bag will give the soup a clearer texture.

Making braised beef

1) Put all the spices in a dry wok with medium heat until fragrant. Set aside.

2) Blanch the beef and the bone. Set aside. Heat oil in a wok with medium heat. Stir-fry the garlic and ginger until fragrant. Add two types of chilli bean sauce. Stir-fry until fragrant.

3) Add the beef. Stir-fry until even. Drizzle wine along the edge of the wok. Stir-fry briefly. Add the soy sauce and dark soy sauce. Stir-fry until fragrant. Add the radish and carrot. Stir well.

4) Add 2.5 liters water in. Put all the spices, leeks, orange and onion into a food filter bag. Put the bag into the wok. Add the rock sugar. Braise over low heat for 2 hours. Add a little salt and sugar if necessary.

5) Put cooked the ramen in a bowl. Add the hot soup. Put some beef and radish on top. Sprinkle with chopped coriander and spring onion. Serve.

Handmade Ramen

1) Combine the flour with the salt and oil. Add the water slowly and knead to form a smooth and shiny dough. Cover with cling wrap for 1 hour.

2) Spread some flour on table. Flatten the dough. Dust a little flour on both sides of dough. Put the dough in noodle cutter to cut into strands (if no noodle cutter, cut dough with knife and hand-pull the noodle strands).

3) Cook the noodles in boiling water for 2 to 3 minutes until done. Rinse with the drinkable cold water briefly and drain. Set aside.

吃得辣可全用郫縣辣豆瓣醬去做，再加 1、2 隻指天椒去調辣。而我的版本是比較正常不太辣的。

For those who like it spicy, all the chilli bean sauce should be replace by the Sichuan spicy chilli bean sauce and add 1 or 2 fresh red chilli. My version is not as spicy.

牛腰你可以整塊放下去，食時再拿出切片；或一早切塊去煮，隨你喜歡。

The beef shank can either be cooked whole and sliced upon serving or cut into pieces before the cooking, according to personal preference.

Chinese Delicacy
中式滋味

糖醋炸蛋

Sweet and Sour Deep-fried Egg with Tomato and Cucumber

最平的食材往往也能做出絕色的美食，
這道菜便是其中之一，
而且大人小朋友也會愛上，
賣相吸睛之餘又好下飯。

The cheapest ingredients can often be made into delicious dishes.
This is one such example.
The attractive presentation and great taste will make both the adults and children fall in love with it.

材料（2-4 人份量）

雞蛋 4 隻
番茄 1 個（去籽切粒）
青瓜 100 克（切粒）

Ingredients (Serves 2-4)

4 eggs
1 tomato (seeded and diced)
100g cucumber (diced)

醬汁料

水 100 毫升
鎮江醋 3 湯匙
茄汁 1.5 湯匙
糖 3、4 茶匙
鹽 1/4 茶匙
生粉 1 湯匙

Seasoning

100ml water
3 tablespoons Zhenjiang vinegar
1.5 tablespoons tomato sauce
3 or 4 teaspoons sugar
1/4 teaspoon salt
1 tablespoon potato Starch

不喜歡用青瓜可用青紅椒代替。

Feel free to replace cucumber with red and green bell pepper.

炸蛋的生熟程度視乎個人喜歡，個人認為這道菜蛋黃炸至剛外表凝固還有輕微流心便最好。

The level of doneness for the deep-fried egg depends on individual preference. Personally I would like to see the outer layer of egg yolk is set yet the centre is custard like.

做法

1) 中火下多點油至油溫 160℃，雞蛋直接一隻一隻打入熱油中去炸。一面先炸 1、2 分鐘。
2) 再反另一面多炸 1 分鐘，至兩面金黃便可。拿起放在碟上隔油，備用。
3) 把所有醬汁料倒進器皿攪勻。
4) 原鑊中火下少許油，先下青瓜略炒一會，再下番茄略炒。加入混好了的醬汁，轉中大火煮約 1 分鐘。再下生粉水埋芡，最後把這個汁料淋在炸好的蛋上便可。

Steps

1) Heat some oil over medium heat to 160 °C. Crack the eggs one by one into the hot oil. Deep-fry one side for 1 to 2 minutes.
2) Turn the fried eggs over and deep-fry the other side for 1 more minute until both sides become golden. Dish up and drain excess oil. Set aside.
3) Combine all sauce ingredients in a container.
4) Heat the wok over medium heat. Add a little oil. Stir-fry the cucumber briefly. Add the tomato and stir-fry briefly. Add the mixed sauce. Adjust to medium high heat and cook for 1 minute. Add the potato starch with water to thicken the sauce. Drizzle sauce over the fried eggs. Serve.

炸物在最後的 30 秒要用大火把油份迫出。

During the deep-frying, adjust to high heat at the last 30 seconds to release the oil in the ingredients.

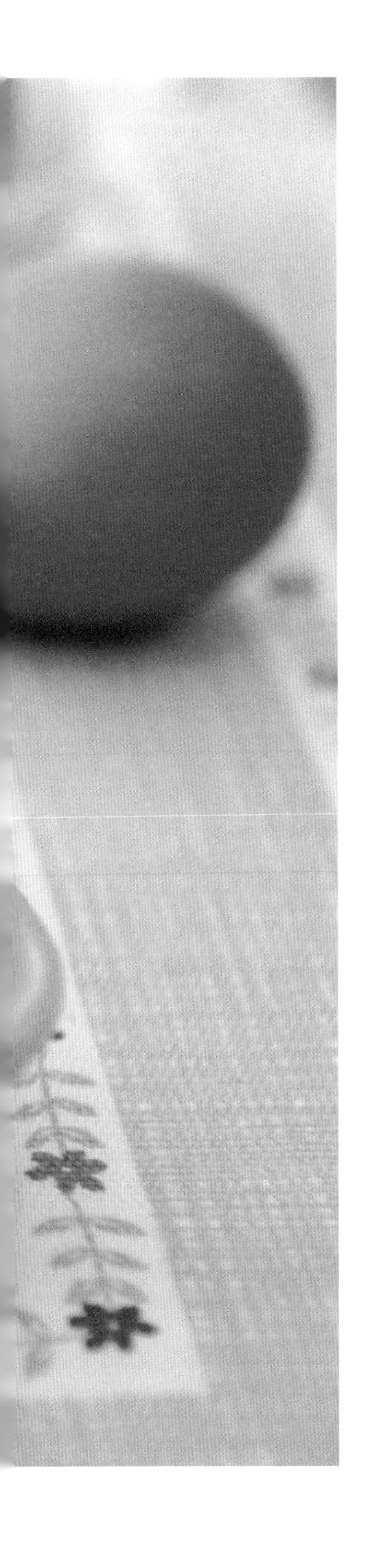

Chinese Delicacy
中式滋味

魚子醬 煙燻鴨蛋

Smoked Duck Eggs with Caviar

本人可算是香港第一代教人玩慢煮的，所以上一本食譜也有些慢煮食譜及資料。然而今期潮流玩「煙燻」，很多米芝蓮餐廳大廚也加入了煙燻菜式，我自己也買了部小型煙燻機玩玩，簡單菜式加上煙燻也即時名貴多了。

I can be regarded as Hong Kong's first generation of food blogger to teach the "Sous Vide" (slow cooking). As such the previous cookbook contains some "slow cook" recipes and information. "Smoked" is the hot new trend recentlty. Even Michelin star chefs are adding smoked cuisine to their menus. I have bought a mini smoker to add that refined touch to some simple dishes.

材料（2-4 人份量）

鴨蛋 4 隻

調味料

豉油 3 湯匙
老抽 5 湯匙
水 250 毫升
味啉 2 湯匙
糖 2 茶匙

若沒有煙燻機，可在鑊上鋪上錫紙，再鋪上煙燻木材加黃糖或米加茶葉加碎冰糖，上面再放個蒸架，放食材在架上蓋上蓋開大火，見出煙後再轉小火去做煙燻。

If smoker is not available, line wok with aluminum foil, add the wood chips for smoking, add the brown sugar or mixture of tea leaf and crushed rock sugar, place a steaming rack on top, place the food on the rack, cover wok and heat with high heat. When the smoke is coming out, adjust to low heat to start the smoking process.

Ingredients (Serves 2-4)

4 duck eggs

Seasonings

3 tablespoons soy sauce
5 tablespoons dark soy sauce
250ml water
2 tablespoons mirin
2 teaspoons sugar

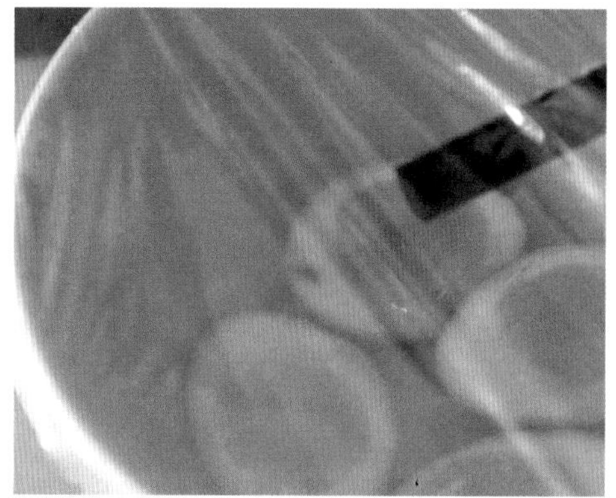

做法

1) 鴨蛋放入凍水開火，水滾後煮 4 分鐘熄火，煮好的鴨蛋放入冰水浸涼，去殼。
2) 所有調味料攪勻，放入鴨蛋浸 1 小時以上。之後取出後放入一器皿，用保鮮紙蓋上。
3) 煙燻機先放入木碎，開機、再點火。有煙出時駁喉入器皿中，注滿煙後抽出氣喉再蓋保鮮紙封着。
4) 煙燻 10 分鐘，鴨蛋開邊加上少許魚子醬或松露醬即可。

Steps

1) Put the duck eggs in cold water. Bring to boil. Simmer for 4 minutes. Turn heat off. Soak the cooked duck eggs in ice water to bring down the temperature. Remove the egg shell.
2) Soak the duck eggs in seasoning ingredients for at least an hour. Dish up and put into a container, cover with cling wrap.
3) Add the wood chips into the smoker. Switch on the smoker and light the fire. When smoke is coming out, connect the hose to the container with the duck eggs. Fill it with smoke. Remove the hose and cover with cling wrap.
4) Stand for 10 minutes. Cut the duck eggs into halves. Add some caviar or truffle sauce on top. Serve.

浸蛋愈久愈入味，其間不時轉動令其上色平均。

The longer the duck eggs are soaked in the seasoning, the better flavors are absorbed. Turn the eggs around and then to ensure even browning.

高湯碗仔翅

Imitation Shark Fin Soup in Hong Kong Style

這可以說是香港的代表性街頭小吃，
我自己也是一個粉絲。
可惜像樣的碗仔翅買少見少，
還是自己在家做吧！

It can be regarded as a signature street food of Hong Kong. I am a big fans too. It is a pity that one that is well made is rarely found nowadays. Let's make it at home.

材料（6-8 碗份量）

雞 1 隻
瘦肉半斤
金華火腿 20 克
木耳 5 片
（浸軟切絲）
冬菇 5 朵
（浸軟切絲）
素翅半包
雞蛋 2 隻（打勻）

Ingredients (6-8 bowls)

1 chicken
1/2 catty lean pork
20g Jinhua ham
5 pieces black fungus
(soak until tender and cut into thin shreds)
5 Chinese mushrooms
(soak until tender and cut into thin shreds)
1/2 pack vegetarian shark fin
2 eggs (beaten)

Seasonings

2 tablespoons soy sauce
3 tablespoons dark soy sauce
3 tablespoons oyster sauce
3 teaspoons white pepper powder
2 teaspoons sugar
4 tablespoons potato starch
Pinch of salt

調味料

生抽 2 湯匙
老抽 3 湯匙
蠔油 3 湯匙
白胡椒粉 3 茶匙
糖 2 茶匙
生粉 4 湯匙
鹽少許

做法

1) 瘦肉出水，雞及素翅用熱水過一下，備用。
2) 湯煲下 3 公升水，放入雞、瘦肉及金華火腿，先中大火煲半小時，再轉小火煮 1 小時。
3) 拿出雞、瘦肉及金華火腿，全部用手撕幼絲，再連同木耳、冬菇及素翅小火多煮半小時。
4) 放入所有調味料，再邊攪伴邊加入蛋汁，若不夠鹹可加入鹽調味，最後打個生粉水芡令湯汁厚身即成。

Steps

1) Blanch the lean pork, and quick blanch the chicken and the imitate shark fin. Set aside.
2) Add 3 liters of water to a pot. Add the chicken, pork and Jinhua ham. Cook over medium high heat for 1/2 hour. Adjust to low heat and simmer for 1 hour.
3) Remove the chicken, pork and Jinhua ham. Tear all cooked ingredients into thin shreds by hands and put it all back in the broth. Add the black fungus, Chinese mushroom and imitate shark fin. Cook over low heat for 1/2 hour.
4) Add all the seasoning ingredients. Keep stirring while adding beaten egg. Add a little salt if necessary. Lastly stir in the potato starch mixture with water to thicken the soup. Serve.

可省卻一些步驟去做個簡易版碗仔翅：用現成雞湯去取代煮高湯，或泰國粉絲代替素翅也可。

To make a simplified version, use the store-bought chicken broth to replace the home-cooked broth or Thai vermicelli instead of vegetarian shark fin.

古法燜乾鮑配花膠

Traditional Braised Dried Abalone wirh Fish Maw

我自己最拿手的其中一道菜式，以前要花一星期才做出的，
現在用壓力煲的話，1 至 2 天便完成。
而自己最近亦愛上了南非養殖的吉品鮑，
一隻 20 頭乾鮑比起日本的平上七成有多。
豪華菜式用自己方法在家做，貴價也可以食得便宜得多。

This is one of the dishes I can cook best. It used to take a week to make it.
Nowadays I use the pressure cooker and it only takes 1 or 2 days to get it done.
Recently I also fell in love with the South African cultured Jipin abalone.
A 20-head dried abalone is cheaper by more than 70% compared to Japan's.
Luxury cuisine could become a lot more affordable when it is home cooked.

材料 (2-4 人份量)

花膠 1 斤（約 5 件）

乾鮑 9 隻

（多少隻隨你喜歡，可以平放在排骨上便可）

排骨 1 斤（斬件）

老雞 1 隻（一開四份）

金華火腿 3 至 5 片（視乎要有多鹹）

Ingredients (Serves 2-4)

1 catty fish maw (about 5 pieces)

9 dried abalones

(quantity according to personal preference, just enough for putting on top of the pork spare rib)

1 catty pork spare rib (chopped into pieces)

1 aged chicken (quartered)

3 to 5 slices Jinhua ham (according to saltiness)

調味料

蠔油 1 湯匙

老抽 1 湯匙（調色用）

冰糖 2 茶匙

Seasonings

1 tablespoon oyster sauce

1 tablespoon dark soy sauce (for coloring purpose)

2 teaspoons rock sugar

正確發花膠方法

1) 原隻花膠不用清洗，直接蒸 30-45 分鐘。此舉可殺死生曬過程中滋生的細菌或微生物，以防在浸發過程中令花膠變壞。
2) 蒸完後用水略為沖洗，再浸冰水放入雪櫃浸過夜。
3) 煮熱一鍋水，放入花膠，待水沸後馬上關火蓋上蓋焗一夜。
4) 把焗好的花膠放在自來水喉下，用最緩慢的水流沖洗半小時。然後將花膠浸進冰水，在雪櫃放置一晚。
5) 隔天看到花膠已充分發脹，便可開始焖煮。如果怕腥的話，可在焖煮前先用熱水加少許米酒（毋須薑葱），把花膠略汆水 1-2 分鐘，便可去腥。
6) 若在步驟 5 發現花膠未發透，請重複步驟 3 及 4。通常來説，普通花膠只需進行以上步驟一次，至於大花膠筒約要用熱水焗兩次，大鰵肚公則要用熱水焗 3、4 次。

注意事項

1) 花膠一遇油便會變「瀉身」（容易溶解），不能再發起。
2) 焖之前要確保花膠已發透。花膠不宜焖煮太久，時間最好控制在 10-20 分鐘（視乎花膠大小）。千萬不要煮超過半小時，否則花膠會溶解及縮小。有些師傅為避免花膠被煮溶，更會選擇使用冷高湯，將花膠浸 2 小時至入味。到上桌之前，才蒸熱花膠及淋上由熱高湯做成的芡汁。
3) 想保存已浸發好的花膠，可將之浸在加了少許鹽的冰水內，然後放進雪櫃。期間要每天換水，最多可儲存 10 天。盡量不要存放在冰格，以免破壞花膠質感。

How to rehydrate the fish maw

1) Before re-hydrating the dried fish maw, steam it for 30-45 minutes (keep it dry before steaming) to kill the bacteria or micro-organisms bred during the drying process.

2) Rinse the steamed the fish maw. Soak it in ice water and keep it in the fridge for one night.

3) Heat up a pot of water and add the fish maw. Bring the water to the boil and turn off the heat immediately. Leave the pot covered and let sit overnight.

4) Place the fish maw under running water and gently wash it for half an hour. Then soak it in ice water again and keep it in the fridge for another night.

5) If the fish maw is fully hydrated the following day, you can prepare it for cooking. To remove the fishy smell, you can boil some water, put in the fish maw and add some amount of Chinese rice wine (no need to add ginger and spring onion) and cook for 1-2 minutes.

6) Repeat step 3 & 4 if the fish maw is not yet fully hydrated. Generally speaking, dried fish maw of normal size only needs the above steps one time. For a large fish maw, you may need to soak it in hot water twice. For a large Pollock maw, you will need to repeat the steps 3-4 times.

Remarks

1) The fish maw melts easily after cooking with oil.

2) Make sure the fish maw is fully rehydrated before stewing. The cooking time is best to keep within 10-20 minutes (depending on its size), or else it will melt and contract. To avoid this problem, some Chinese chefs will soak it in a cold stock for 2 hours to enrich its flavour, then reheat it by steaming and pouring heated stock over it before serving.

3) To store the hydrated fish maw, you can soak it in ice water mixed with a small amount of salt and keep it in the fridge. You can keep it in the fridge up to 10 days. Remember to change the water every day. Avoid putting it into the freezer as the cold temperature will ruin the texture of the fish maw.

乾鮑浸發過程

1) 先將乾鮑用水浸過，水要蓋過鮑魚，放入雪櫃用保鮮紙或食物盒蓋好，每日換水，連續浸三日。
2) 冰水浸過幾天後，用煲加水後開火，水的份量要能蓋過鮑魚便夠。水滾後放入鮑魚蓋上蓋，水再滾起後即熄火，之後焗水過夜。
3) 焗過水的鮑魚取出洗淨，盡量不要弄破鮑魚本身。再斜斜剪開鮑魚咀，用牙籤輕輕將內裏的腸臟及污穢取出，洗淨，按一按鮑魚中間，若夠彈性即發透。若不夠可再浸冰水放入雪櫃多一晚，至發透為止。

How to rehydrate dried abalone

1) Soak dried abalones in water. The water should cover the abalones. Cover with cling wrap or put into a covered container and chill in fridge. Change water daily and soak for 3 days.
2) After soaking is done, bring some water to boil in a pot. The amount of water should be enough to cover the dried abalones. Add dried abalones to the boiling water. Cover pot. Bring to boil. Turn heat off. Let sit overnight.
3) Rinse well the abalones. Be careful not to tear the abalones. Use scissors to cut open the abalone mouth aslant. Use toothpick to gently discard the intestines. Rinse well abalones. Firmly press the center of each abalone. The soaking is done if the texture is springy. If it is not fully done yet, soak the abalone in ice water and chill in fridge overnight until the abalone is fully soaked through.

注意事項

乾鮑浸發若不夠透徹，會很難入味，乾鮑亦千萬不可用薑、葱或酒去焗水，因會破壞乾鮑本身的鮮味，浸發時間亦以鮑魚大小及種類而定，25-30 頭最少要發 3 至 4 天，鮑魚愈大則要多加 1 至 3 天去發。

Remarks

If dried abalone is not fully soaked, it will not be able to absorb the flavours well. Do not soak the dried abalone with ginger, spring onion or wine in boiling water as this will destroy the taste of abalone. The soaking time varies according to the size and the type of dried abalones. A 25-30 head dried abalone needs to be soaked for 3 to 4 days. The bigger ones will need to add another 1 to 3 days.

做法

1) 排骨出水後抹乾，再半煎炸至金黃，放在廚房紙吸去多餘的油份。
2) 在煲底放入竹笪防黐底，先鋪上煎過的排骨。
3) 在排骨上平鋪上鮑魚，再在鮑魚上鋪上出過水的老雞，可加入之前浸鮑魚的水或普通水至剛蓋鮑魚 1 吋便夠。

Steps

1) Blanch the spareribs, drain and dry. Quick blanch the aged chicken, set aside. Deep fried spareribs until golden. Use the kitchen tissue to absorb excess oil.
2) Put a bamboo rack in the bottom of the pot. Place the fried spareribs as the first layer.
3) Place the abalones as the second layer on top of the pork rib. Place the aged chicken pieces on top of abalones. Add some water used for soaking the dried abalones or plain water until water level is 1-inch above the abalones.

20 頭以下鮑魚燜 10 至 14 小時便可，20 頭以上燜 15 至 18 小時，10 頭以上要燜 18 至 22 小時。

Abalone that is below 20-head needs about 10 to 14 hours of braising. Those above 20-head will need 15-18 hours whereas those above 10-head will need 18 to 22 hours.

4) 水滾後轉慢火燜 3 小時，後熄火焗至冷，再加水開火再燜 3 小時，再熄火焗過夜，重複燜焗步驟每天2-3次，大約共 12-15 小時。之後用牙籤嘗試能否輕易穿過鮑魚中間，能輕易穿過為之腍，否則請加水再燜至腍身。

5) 燜煮其間注意不夠水便要加水但又不要加太多，水剛蓋過鮑魚少許便可。否剛令湯汁淡味而令鮑魚味道不夠濃郁，而在最後燜的 3 小時才加入金華火腿去調味。

4) Bring water to boil. Adjust to low heat and braise for 3 hours. Turn heat off and let stand (pot covered) until water has cooled off. Add water. Turn on the heat and braise for another 3 hours. Turn heat off and let stand overnight. Repeat the braising and cooling off process for 2 to 3 times each day, totaling 12-15 hours. Use a toothpick to test whether it can easily poke through the abalone. The tenderness is just right if the toothpick can easily poke through. Add some water and repeat the braising if the texture is not there yet.

5) Pay attention to the water level during the braising. Add some, but not too much, if the water is not enough. Add just enough to cover the abalones. Too much will cause the sauce to taste bland and dilute the rich flavor of abalone. Add Jinhua ham at the final 3 hours of the braising.

千萬不要下太早下金華火腿或下鹽燜煮，因「鹹」會令鮑魚變硬不易燜至腍身。

Do not add Jinhua ham or salt too early during the braising because "saltiness" will toughen up the abalone texture

6) 燜好的鮑魚，把鮑汁及鮑魚拿起備用。拿出的鮑魚，放在陰涼的地方風乾 30 分鐘或更多，此舉可令鮑魚轉溏心色。
7) 剩下的鮑汁加入調味料滾起，放入花膠後熄火焗 15 分鐘或開小火燜 5-10 分鐘。再將鮑魚回鍋弄熱，最後打個生粉水埋芡即成。

6) Dish up the cooked abalones, strain the soup through a sieve and set aside. Put abalones in a cool place and let them rest for 30 minutes or more. This is to let the abalones seal in the juices and become nicely browned.
7) Add some seasoning to the remaining abalone soup. Bring to boil. Add fish maw. Turn heat off and let stand for 15 minutes with lid closed or braise over low heat for 5-10 minutes. Return abalone to the pot. Whip up a little potato starch solution to thicken the sauce. Serve.

燜完鮑魚剩下湯料不要浪費，我自己會加水再煲出高湯。

Do not discard the soup leftover from the braising. I will add some water to re-boil it into stock.

煮好的鮑魚，可每隻獨立用保鮮紙包好，再用乾毛巾包好放入冰箱。鮑汁亦可分開放入冰箱，食時從冰箱拿出解凍再翻熱打芡便可。

Wrap up the leftover abalone separately with cling wrap and wrap all again with dry towel before putting into fridge. Set aside the abalone sauce and keep in fridge as well. Remove from fridge, let thaw, reheat and add a little corn starch solution before serving.

花膠若發得夠透是不用煮很久的，愈煮得耐花膠愈「瀉」身。

Fish maw that has been soaked through does not require for long cooking time. It will dissolve when over-cooked.

若夏天煮這個菜式，要在盡量保持着溫度，因放置太久會容易變壞，建議睡前先煲熱才過夜，早上起來也煲熱一下。

If this dish is cooked in the summer, try to maintain the temperature as the dish will become spoilt if it is set aside for too long. As a precaution, heat it before going to bed and reheat it in the morning.

若用壓力煲，燜的時間可以大大減少。用壓力煲十分鐘等如普通煲 1 小時，所以要燜 15 小時的鮑魚，我會用壓力煲燜 2 小時，再由它焗過夜讓鮑魚入味。翌日再加入金華火腿多燜 1 小時便可。

A pressure cooker will greatly shorten the cooking time. 10 minutes with pressure cooker is equivalent to 1 hour using a regular pot. Handling abalone that requires 15 hours of braising, I will braise with pressure cooker for 2 hours and leave it overnight to let abalones absorb the flavour. Add some Jinhua ham the following day and braise for another hour will do.

富貴盆菜

Big Bowl Feast (Poon Choi)

盆菜可是近年最受歡的新春菜式之一，記得中學時因為入讀了一間在上水圍村學校，所以每年都會舉辦大型的「食山頭」（即是掃墓食盆菜）活動。全校學生一起行山祭祖，之後會直接在山上吃盆菜。記得當時的盆菜並不是太好吃，沒有火爐烘着，草地鋪上膠枱布便坐地而吃。裏面都是些普通材料如切雞、燒肉、燜五花腩、豬皮、魚蛋、蘿蔔等……味道並不是十分吸引。但多年來經過改良，愈吃愈富貴，演變成現在的名貴菜式，味道亦愈來愈講究。跟傳統的真的有很大分別啊！

“Poon choi”is fast becoming one of the most popular New Year dishes in recent years. I could still remember that the high school in Sheung Shui Wall Village where I was studying, would hold a large-scale "Eat Shantou" activity (that is a Poon Choi served during Ancestral tomb-sweeping day) every year. The whole school would hike to the mountain to pray to the ancestors, and would eat the casserole dish on the mountain. The Poon Choi at that time was not as delicious. There was no stove to keep the dish hot and everyone would sit on the plastic tablecloth placed on the grass. There were common ingredients such as chicken, roast pork, braised pork belly, pig’s skin, fish ball and radish etc. The taste is average.

However, after years of improvement, this dish has become more and more costly, evolving into the current version of luxury dishes. The taste has also become more refined, a vast difference from the traditional ones.

材料（10 人份量）	Ingredients (Serves 10)
金蠔或蠔豉 10 隻	10 dried oysters or golden dried oysters
罐頭鮑魚 10 粒裝 1 罐	1 can of abalone (10 abalones)
已發花膠約 4 大片	4 big pieces rehydrated fish maw
厚身冬菇 10 朵（浸軟）	10 thick-fleshed Chinese mushrooms (soaked until tender)
豬腩肉 1 斤半（切塊出水）	1.5 catty pork belly (cut into pieces and blanch)
枝竹 1 斤	1 catty dried tofu stick
豆卜約 20 多粒	20 + tofu puffs
豬皮 1 斤（出水）	1 catty pig skin (cut into pieces and blanch)
水魷 1 斤（切塊去衣出水）	1 catty squid (cut into pieces, peeled and blanch)
白蘿蔔 2 斤（切塊）	2 catties radish (cut into pieces)
西蘭花 2 大棵（切塊）	2 stalks broccoli (cut into florets)
大蝦 10 隻	10 big prawns
炸魚蛋約 20 多粒	20 fried fish balls
燒鴨半隻（現成斬件）	1/2 roast goose (store-bought with chopped)
白切雞半隻（現成斬件）	1/2 poached chicken (store-bought with chopped)
陳皮 1 塊	1 piece dried tangerine peel
桂皮 1 枝	1 cinnamon stick
八角 2 粒	2 pieces star anise
蒜頭 1 個（切碎）	1 bulb garlic (finely chopped)
乾葱 2 粒（切碎）	2 cloves shallot (finely chopped)
薑片 8 片	8 slices ginger
葱段 3 棵	3 stalks spring onion (sectioned)

調味料 Seasonings

調味料	Seasonings
南乳 1 磚	1 piece preserved red tofu
磨豉醬 1 湯匙	1 tablespoon ground bean paste
冰糖 4 湯匙	4 tablespoons rock sugar
老抽 4 湯匙	4 tablespoons dark soy sauce
蠔油 4 湯匙	4 tablespoons oyster sauce
糖 3 茶匙	3 teaspoons sugar
雞湯 400 毫升	400ml chicken broth
金華火腿雞上湯 400 毫升	400ml Jinhua ham chicken broth
紹興酒 2 湯匙	2 tablespoon Shaoxing wine
麻油半湯匙	1/2 tablespoon sesame oil
鹽 1 茶匙	1 teaspoon salt
美極鮮醬油 2 湯匙	2 tablespoons Maggi Seasoning

金華火腿雞湯可用現成的或自己用雞湯加金華火腿。

Jinhua ham chicken broth can be either the store-bought version or made with chicken broth cooked with some Jinhua ham.

豬皮買回來後浸水，去毛，再出水。

Soak the pig skin in water, remove the hair and blanch.

做法

食材要分別處理煮好，最後再砌在一起；其實這裏分別有幾道菜的食譜。食譜如下：

Steps

The ingredients should be prepared and cooked separately and finally put together; in fact, this dish is divided into several recipes:

南乳枝竹燜豬腩肉

1) 豬腩肉切塊，出水抹乾，再走油。
2) 南乳加入 4 湯匙水壓爛成糊，備用。
3) 下油，下 4 片薑、蒜蓉、乾葱、葱段炒香，再下南乳及磨豉醬炒香。
4) 加入豬腩肉略炒，再加入 2 公升水、陳皮、桂皮、八角、2 湯匙冰糖、1 湯匙老抽，水滾後轉細火燜 1 小時。
5) 最後加入枝竹、豆卜，再燜 10 分鐘，即成，備用。

Preserved Tofu Braised Pork Belly and Tofu Stick

1) Dry up the blanched pork, quick deep fried in the hot oil briefly. Set aside
2) Mash preserved the red tofu with 4 tablespoons water and mixed well. Set aside.
3) Heat the oil. Stir-fry 4 slices ginger, garlic, shallot and spring onion until fragrant. Add preserved red tofu and ground bean paste, stir-fry until fragrant.
4) Add the pork and stir-fry briefly. Add 2 liters of water, dried tangerine peel, cinnamon stick, star anise, 2 tablespoons of rock sugar and 1 tablespoon dark soy sauce. Bring water to boil. Adjust to low heat and braise for 1 hour.
5) Add the tofu stick, tofu puff and braise for another 10 minutes. Serve.

蘿蔔豬皮魷魚魚蛋

1) 下油，下蒜蓉及乾葱炒香，加入蘿蔔略炒。
2) 再加入400毫升雞湯，老抽、蠔油各2湯匙，糖2茶匙，滾起轉小火燜 15 分鐘。
3) 再加入已出水的豬皮及炸魚蛋，多燜 10 分鐘。
4) 魷魚在最後放入多煮一會至熟便可。

Stewed Pig Skin, Squid and Fish Ball with Radish

1) Heat a little oil in wok. Stir-fry the garlic and shallot until fragrant. Add the radish and stir-fry briefly.
2) Add 400ml chicken broth, 2 tablespoons each of dark soy sauce and oyster sauce, 2 teaspoons sugar. Bring to boil. Adjust to low heat and braise for 15 minutes.
3) Add the pig skin and fried fish ball. Braise for another 10 minutes.
4) Lastly add the squid. Cook for a while until done. Set aside.

花膠鮑魚燜冬菇

1) 下油，再下幾片薑及葱段炒香，加入已浸軟的冬菇略炒，圍邊加入 1 湯匙紹興酒略炒。
2) 加入整罐罐頭鮑魚（連水），再加約 400 毫升金華火腿雞上湯。加入蠔油 1.5 湯匙、1 湯匙冰糖及 1 湯匙老抽。滾後轉小火燜 30 分鐘。
3) 加入花膠件，再燜約 10 分鐘至腍身便可，最後用生粉水埋芡即成，備用。

Braised Fish Maw and Abalone with Chinese Mushroom

1) Heat a little oil in wok. Add a few slices of ginger and spring onion. Stir-fry until fragrant. Add the soaked Chinese mushrooms and stir-fry briefly. Drizzle 1 tablespoon of Shaoxing wine along the edge of wok. Stir-fry briefly.
2) Add the whole can of abalone (including the juice). Add 400ml Jinhua ham chicken stock. Add 1.5 tablespoons oyster sauce, 1 tablespoon rock sugar, 1 tablespoon dark soy sauce. Bring to boil. Adjust to low heat and braise for 30 minutes.
3) Add the fish maw. Braise for another 10 minutes until tender. Add a little potato starch solution to thicken the sauce. Set aside.

金蠔

洗淨抹乾，加入 1 湯匙紹興酒、半湯匙麻油、1 茶匙糖、鹽及生粉醃一會。之後蒸 15 分鐘即成。

Golden Oyster

Rinse and pat dry the dried oysters. Marinate with 1 tablespoon Shaoxing wine, 1/2 tablespoon sesame oil, 1 teaspoon sugar, salt and potato starch. Let stand for a while. Steam for 15 minutes. Set aside.

大蝦

下油，把蝦煎香至八成熟，倒入 2 湯匙美極醬油，再略煎至色澤平均便可。

Big Prawns

Heat a little oil in wok. Pan-fry the big prawns until 80% done and fragrant. Add 2 tablespoons Maggi Seasoning. Pan-fry briefly until the prawns are evenly browned. Set aside.

西蘭花

可灼熟或蒸 3 分鐘便可，備用。

Broccoli

Blanch the broccoli until done or steam broccoli for 3 minutes. Set aside.

砌盤菜

1) 先放入蘿蔔在最底一層，加少許汁；
2) 第二層再加入枝竹豆卜豬腩肉及少許汁；
3) 第三層鋪上魷魚豬皮魚蛋；
4) 第四層鋪上切雞及燒鴨；
5) 第五層先圍邊鋪上西蘭花；
6) 再在裏面鋪上大蝦、金蠔、鮑魚、花膠，中間鋪冬菇。最後把燜鮑魚芡汁全淋上即成。

Assembling casserole

1) Add the radish and pig skin and the bottom. Add a little sauce.
2) Add a layer of tofu stick, tofu puff, pork and a little sauce.
3) Add a layer of squid, pig skin and fish ball.
4) Add a layer of poached chicken and roast duck.
5) Place a ring of broccoli at the top most layer.
6) In the broccoli ring, arrange big prawns, golden oysters, abalones and fish maw. Place Chinese mushrooms at the center. Lastly drizzle braised abalone sauce over the ingredients. Serve.

Asian Gourmet

亞洲風味

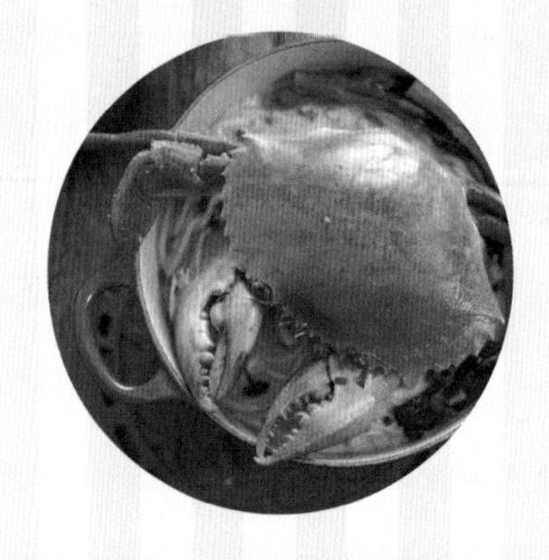

泰式生蝦

Raw Prawns in Thai Style (Goong Chae Nam Pla)

有人常問，哪裏來的海鮮可以生吃？為甚麼日本大部份海鮮也可做刺生，但其他地區的好像卻沒有……其實，所有海鮮是從大海捉上來的都可以生吃！但最重要的是要看海域的水質如何？由海上到海鮮檔的水質有否被污染？找一個可靠的來源便可以，當然自己也要小心評估風險，而蝦是比較安全做刺生的海鮮之一，不妨一試。

People often asked, “seafood from which place can be eaten in raw? Why most seafood from Japan can be made into sashimi style but those from other areas are not so?” The truth is, most seafood freshly caught from the sea can be eaten in raw. The important factors are the water quality in these coastal areas and the risk of pollution while being transported from the sea to the store. What you need is a reliable source and not taking for granted the risk evaluation. Prawn is one of the safer seafood that can be made into sashimi, worth to try.

材料（4人份量）

活海蝦 10-12 隻

Ingredients (Serves 4)

10-12 live sea prawns

泰式汁料

青辣椒 2 隻（喜歡辣可以多加 1、2 隻）
芫荽 1 小棵（連根洗淨）
蒜頭 3 瓣

Thai style sauce

2 green chillis (add 1 or 2 more for those who like it spicy)
1 coriander (rinse well, including the root)
3 cloves garlic

調味料

魚露 2 湯匙
青檸汁 2.5 湯匙
糖 2 茶匙

Seasonings

2 tablespoons fish sauce
2.5 tablespoons lime juice
2 teaspoons sugar

做法

1) 將蒜頭、辣椒、芫荽用攪拌機打爛，或用刀啄成蓉。加入調味料，攪勻，放入雪櫃備用。
2) 蝦用食水洗淨，切頭。去殼，殼留蝦尾最後一節。由背部開邊。挑去中間的腸浸入冰食水一會令蝦肉爽身。
3) 把蝦拿出用廚房紙印乾放上碟，用保鮮紙包好，放入雪櫃雪凍至食時才拿出。通常雪最少半小時以上，蝦肉才會更爽身，配上汁醬，即成！

Steps

1) Mince the garlic, chilli and coriander with blender or knife. Add the seasoning ingredients. Stir them well. Chill in fridge.
2) Rinse the prawns with drinking water. Remove the head and shell of prawns. Retain the last segment of the shell of prawn tail. Slit back of prawns. Remove the intestine. Soak the prawns in ice drinking water to enhance the crunchiness of prawn meat.
3) Pat dry the prawns. Arrange them in the plate. Wrap the plate with cling wrap. Chill in fridge for at least half an hour to get the crunchier texture. Remove from fridge upon serving. Eat with sauce.

最好用草蝦去做，肉質比較爽身。

Tiger prawn is ideal for this dish. The meat texture is crunchier.

這汁醬可多做一點放入雪櫃存放並隨時享用，大約可存放一星期。

Feel free to make more of the sauce to be kept in fridge for future use. Can be kept up to 1 week.

泰式炒金邊粉

Pad Thai

這是一個大部份人去泰國餐廳都會叫的菜式，在家做「易過借火」。很多泰國調味料也可以在九龍城南角道買到。而且我自己煮蝦喜歡連殼煮，因為鮮味會更香。若怕吃時麻煩可在煮之前先行去殼也可。

This is a dish that most people will order when they eat at a Thai restaurant. It is easy to make at home. Most of the Thai seasoning ingredients are available at Nam Kok Road, Kowloon City. Also, I personally like to cook prawns with shell. The prawn aroma will be richer. However, for easy eating, feel free to peel off the shell before cooking.

材料 (2 人份量)

泰式金邊粉 150 克 （用水浸軟）
中蝦 6 隻 （背部開邊）
豆乾 1 磚 (切條)
雞蛋 1 隻
芽菜 適量
紅尖椒 1 條
葱 4 條
蒜蓉 2 茶匙

調味料

羅望子汁 2 湯匙
椰糖 1 湯匙
辣椒粉 1 茶匙
魚露 2 湯匙

伴碟

青檸半個
花生碎 適量

Ingredients (Serves 2)

150g pad Thai noodles
(soaked in water until tender)
6 medium prawns
(slit back of prawns)
1 brick dried bean curd
(shredded)
1 egg
Some bean sprout
1 red long chilli
4 stalks spring onion
2 teaspoons minced garlic

Seasonings

2 tablespoons tamarind sauce
1 tablespoon coconut sugar
1 teaspoon chilli flake
2 tablespoons fish sauce

Garnishing

1/2 lime
Some crushed peanut

做法

1) 把所有調味料勻和在一起成醬汁，備用。
2) 中大火下油，先放蝦煎香至七成熟，其間下少許鹽調味，拿起備用。
3) 原鑊加少許油，下雞蛋炒碎，要快手及不要過熟。拿起備用。
4) 原鑊加少許油，先下蒜、葱及紅椒炒香，再下豆乾及芽菜略炒一會。
5) 放入金邊粉及調味醬汁炒 1 分鐘，再加入之前的蝦及蛋炒勻，最後再加入少許魚露調味，上碟後加入伴碟食材，即成。

Steps

1) Combine all the seasoning ingredients and set aside.
2) Heat oil in wok over medium high heat. Pan-fried the prawns until fragrant and 70% cooked. Add a pinch of salt as seasoning. Set aside.
3) Add a little oil to the same wok. Stir-fry the egg into small pieces. Stir-fried swiftly and do not over-cook. Set aside.
4) Add a little oil to the same wok. Stir-fry the garlic, spring onion and red chillis until fragrant. Add the dried bean curd and bean sprout. Stir-fry briefly.
5) Add the noodles and seasoning mixture. Stir-fry for 1 minute. Add the prawns and fried egg. Stir-fry until even. Add some fish sauce. Transfer to plate. Add the garnishing. Serve.

金邊粉不能炒太久，否則粉的口感會被破壞。

Do not stir-fry the pad thai for too long. It will destroy the texture of the noodles.

泰國人喜歡用魚露代替鹽去調味，最後隨自己口味加入魚露調至喜歡的鹹度便可。

Thai people prefer to use the fish sauce instead of salt as seasoning. Fish sauce is added at last according to personal preference, to the desired level of saltiness.

泰式香葉包雞

Thai Style Pandan Leaf Deep-frying Chicken

我教烹飪班時，這是一個很愛歡迎的菜式，
吃過後每位同學仔也讚不絕口，
但炸的過程一定要小心彈油。

In my cooking classes,
this is a very popular dish
that bound to win lots of praises
from the students when the cooking is done.
Pay close attention to the deep-frying
to prevent hot oil from splattering.

材料（4 人份量）

雞脾肉 1 隻
（去皮去骨切塊）
斑蘭葉 15 塊
（洗淨抹乾）

Ingredients (Serves 4)

1 piece chicken thigh
(remove skin and bone, cut meat into pieces)
15 Pandan leaves
(rinsed and pat dry)

調味料

豉油 1 湯匙
魚露 2 茶匙
蠔油 2 茶匙
糖 1 茶匙
麻油 1 茶匙
白胡椒粉 1 茶匙

Seasonings

1 tablespoon soy sauce
2 teaspoons fish sauce
2 teaspoons oyster sauce
1 teaspoon sugar
1 teaspoon sesame oil
1 teaspoon ground white pepper

醃料

香茅莖 1 枝（切碎）
芫荽根 2 棵（切碎）
蒜頭 2 瓣（切碎）

Marinade

1 stalk lemongrass (finely chopped)
2 stalks coriander with root
(finely chopped)
2 cloves garlic (finely chopped)

做法

1) 把所有調味料及醃料放入雞件，攪勻。放雪櫃醃 1 小時。
2) 斑蘭葉抹乾淨，底部剪平，再摺成一個三角筒形。
3) 把醃好的雞塊放在斑蘭葉內（三角筒形內）。
4) 再對角一直摺上去，直至葉把雞件包好，葉的尾部收入去。
5) 再用牙籤把尾部斜斜地穿實便可。
6) 下油，當油溫達 180℃，用中火把香葉雞炸 5-7 分鐘，之後放在廚房紙吸油。即成。

4.3

4.4

6

Steps

1) Combine the chicken thigh meat with all seasoning and all marinade ingredients. Chill in fridge marinade for 1 hour.
2) Wipe to clean the pandan leaves. Trim the bottom and fold into triangular cone.
3) Place the marinated chicken meat in the triangular pandan cone.
4) Fold the pandan leaf across to cover all the chicken meat and tuck in the pandan leaf tail.
5) Poke a toothpick at the tail end to hold the leaf together.
6) Heat oil to 180 °C. Deep-fry the pandan leaf chicken over medium heat for 5-7 minutes. Dish up and put on kitchen tissue to absorb the excess oil. Serve.

CHIN-SU

Asian Gourmet
亞洲風味

椰青蛋白蒸蟹

Steamed Crab with Young Coconut and Egg White

椰青配海鮮其實是很合適的，
無論蜆、蝦及蟹，也十分合適。
越南人很喜歡用它做菜式，
今次為大家示範這道簡單大體的家鄉名菜。

Young coconut goes well with seafood,
be it clam, prawn or crab.
Vietnamese people are fond of using young coconut
in creating various dishes.
Let me demonstrate this simple yet presentable
signature dish from my home country.

材料（2-4 人份量）

肉蟹 1 隻
椰青 1 個
蛋白 2 隻

Ingredients (Serves 2-4)

1 meat crab
1 young coconut
2 egg whites

調味料

白砂糖 1 茶匙
魚露 2 湯匙
月桂葉 3 塊

Seasonings

1 teaspoon white caster sugar
2 tablespoons fish sauce
3 bay leaves

斬蟹步驟

1) 先把刀打斜放在蟹蓋與蟹臂的中間位置上，用另一隻手在刀背施力拍一下，切去蟹鉗。
2) 把蟹身翻轉，拆去蟹蓋。
3) 除去蟹肺，再用刷子刷去污垢。
4) 把蟹身對半斬開成 2 份，然後再把每邊切成 3 份。
5) 把前臂與蟹鉗切開，再平刀敲碎蟹鉗，即成。

To prepare the crab

1) Place the knife diagonally between the top shell and the front legs. Pat the back of the knife with another hand to chop off the front legs.
2) Turn the crab over and remove the top shell.
3) Remove the gills and brush away the dirt.
4) Halve the crab then chop each half into 3 pieces.
5) Chop off the claws from the front legs. Crack the claws and front legs shell with the side of the knife.

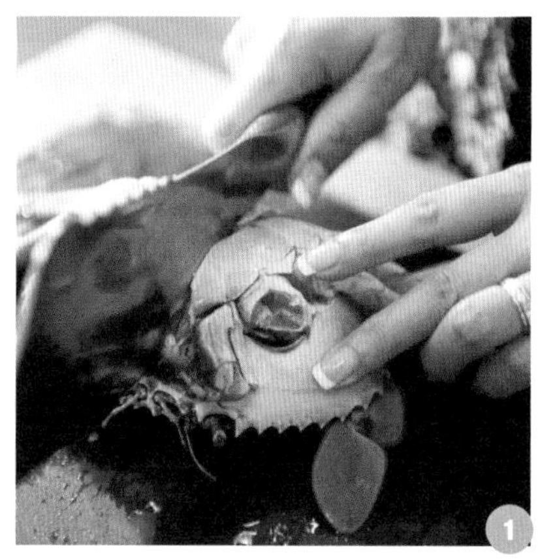

做法

1) 開椰青，然後倒出椰青水再起肉備用。
2) 將椰青水加入蛋白、魚露、糖攪拌均勻。
3) 蟹放上深碟內，加入拌好的蛋白椰青水，再鋪上椰青肉及月桂葉。大火蒸 20 分鐘便完成。

Steps

1) Cut and open the young coconut. Pour out the coconut juice. Scrape the coconut meat.
2) Combine the coconut juice with egg white, fish sauce and sugar.
3) Arrange the crab pieces in a deep plate. Add the coconut juice mixture. Add the coconut meat and the bay leaves. Steam over high heat for 20 minutes. Serve.

蒸時需包上耐熱保鮮紙，防止倒汗水和令蛋白起皺。

Cover with the heat-resistant cling wrap to prevent condensation from dropping onto the steamed egg white causing the egg white to become uneven surface.

越南
豬耳扎肉

Vietnamese Pig Ear Pork Roll (Cha Lua)

從小到大也十分喜歡吃扎肉，
但要做到夠彈性又美味，
有幾個步驟一定要跟着啦！

Pork roll is one of my favorite foods since young.
For it to be tasty and springy,
there are a few must-follow steps.

材料（2條扎肉份量）

瘦免治豬柳肉 1 斤
豬耳 1 隻
木耳 1 片（浸軟切絲）
薑 5 片
葱段 1 棵

調味料

越南魚露 1 湯匙
糖 2 湯匙
鹽 1.5 茶匙
木薯粉 1 湯匙
粟粉 2 湯匙
白胡椒粉 2 茶匙
白胡椒粒 2 茶匙（壓碎）
紹酒 1 湯匙
發粉 1 茶匙
白醋少許

Ingredients (To yield 2 pork rolls)

1 catty lean pork fillet minced
1 pig ear
1 piece black fungus Mu Er (soaked until tender and cut into thin shreds)
5 slices ginger
1 stalk spring onion (sectioned)

Seasonings

1 tablespoon Vietnamese fish sauce
2 tablespoons sugar
1.5 teaspoons salt
1 tablespoon tapioca starch
2 tablespoons corn starch
2 teaspoons ground white peppercorn
2 teaspoons white peppercorn (crushed)
1 tablespoon Shaoxing wine
1 teaspoon baking powder
A little vinegar

做法

1) 免治豬肉加入魚露、糖、鹽、白胡椒粉、木薯粉及粟粉攪勻，放入冰格雪 1.5 小時。要夠凍但又不結冰。
2) 新鮮豬耳用鹽及白醋洗淨，出水後再清洗。放入一煲水，加入薑片、葱段、紹酒、1 茶匙鹽及糖，用慢火煮 1 小時，煮好的豬耳抹乾切絲。
3) 發粉加入 6 湯匙水攪勻備用。
4) 雪凍了的豬肉放入攪拌機內，先用中速攪 2 分鐘，再加入發粉水再攪 2 分鐘至起膠。

豬肉要夠瘦扎肉才會夠挺身，請先去除所有脂肪才攪碎。而且一定要雪凍才會起膠。

The pork has to be lean enough to get the firmer texture. Remove all fat before the minced. The blended meat should be chilled in freezer to make it sticky.

Steps

1) Combine the minced pork with the fish sauce, sugar, salt, ground white pepper, tapioca starch and corn starch. Chill in the freezer for 1.5 hours. Chilled but not frozen.
2) Clean the pig ear with salt and white vinegar. Blanch and rinse well. Put the pig ear in a pot of water. Add the ginger, spring onion, Shaoxing wine, 1 teaspoon salt and sugar. Cook over low heat for 1 hour. Pat dry the cooked pig ear and cut it into thin shreds.
3) Mix the baking powder with 6 tablespoons water.
4) Put the chilled pork in a blender. Blend with medium speed for 2 minutes. Add the baking powder solution and blend for another 2 minutes until sticky.

可用蕉葉代替保鮮紙包紮豬肉去蒸更加正宗。蕉葉用濕布抹乾、塗上少許油便可用。

For a more traditional, wrap the pork rolls with banana leaf instead of cling wrap. Wipe banana leaf with wet cloth, brush with a little oil on top before put the meat.

5) 加入配料豬耳、木耳及白胡椒粒拌勻。
6) 豬肉放在高溫保鮮紙上，包起紮成 2 條腸型。大火蒸 25 分鐘。
7) 蒸熟的扎肉拿開高溫保鮮紙，乘涼後切片即可食用（雪凍後再吃會更加好吃）。

5) Add the pig ear, black fungus and white peppercorn. Mix well.
6) Place the pork mixture on the heat-resistant cling wrap. Wrap to shape of sausage. Steam over high heat for 25 minutes.
7) Remove the cling wrap. Let cool and slice pork roll. Serve. (the taste is even better when it is first chilled in fridge)

做多了的扎肉可包裹好放冰箱，可存放 3 個月。

Excess pork roll can be wrapped up and frozen up to 3 months.

豬肉你可以配搭任何配料、如黑、白胡椒粒等……但所有配料一定要在最後才加入。

Other side ingredients can be added to the pork, such as black or white peppercorn but they have to be added at the final stage.

若做淨扎肉，請省卻步驟 2 及 5 便可，亦不需要準備那些材料。

If to make the plain pork roll, omit steps 2 and 5 as well as those side ingredients.

Asian Gourmet
亞洲風味

越南
生熟牛河

Vietnamese Style Raw and Cooked Beef with Flat Rice Noodles (Pho Bo)

作為半個越南人，
怎可能不把這個越南最有名的菜式做好。
正宗的越南牛河湯底一定要清澈且香濃，
要做到這點真是要一些竅門的，
現在跟大家分享一下。

Being half of a Vietnamese, there is no excuse for me to not have mastered the making of this most famous Vietnamese cuisine. Authentic Vietnamese beef soup must be clear yet rich. To do it right does require some skills that I am happy to share with you in this recipe.

湯底材料（4-6 人份量）

牛腰半斤（整條）
牛骨 3 斤（斬件）
鮮雞 1 隻（去皮）
洋葱 3 個（切塊）
乾葱 6 粒（開邊）
薑 8 大片

香料

丁香 5 粒
八角 5 粒
肉桂 2 枝
草果 1 粒（拍碎只要外殼）
月桂葉 2 片
花椒 2 茶匙
白胡椒粒 1 湯匙
小茴香 1 茶匙

調味料

冰糖 50 克
魚露 6 湯匙
鹽 2 茶匙
糖 1 茶匙

河粉材料

生牛肉 200 克
河粉 1.5 斤
洋葱 1 個（切條）
芫荽葱少許（切碎）
金不換 5 棵
指天椒 3 條（切粒）
青檸 2 個（1 開 4）
芽菜少許

貼士 tips #1

要煮清澈湯底，全程一定要小火開蓋煮，期間不時撇走油及骯髒物。

To have a clear soup base, the entire simmering process must be done over low heat uncover the pot. Remove the oil and impurities every now and then.

Soup Ingredients (Serves 4-6)

1/2 catty beef shank (whole piece)

3 catties beef bone (chop into pieces)

1 fresh chicken (skinned)

3 onions (cut into pieces)

6 cloves shallot (halved)

8 big slices ginger

Spice

5 pieces clove

5 pieces star anise

2 cinnamon sticks

1 piece Cao Guo (smashed, use the shell only)

2 bay leaves

2 teaspoons Sichuan peppercorn

1 tablespoon white peppercorn

1 teaspoon fennel seeds

Seasoning

50g rock sugar

6 tablespoons fish sauce

2 teaspoons salt

1 teaspoon sugar

Ingredients for beef noodle

200g raw beef

1.5 catties flat rice noodle

1 onion (shredded)

A little coriander and spring onion (diced)

5 stalks sweet basil (leaves only)

3 fresh red eye chillis (diced)

2 limes (quartered)

Some bean sprouts

牛骨是主要材料，牛肉出味一定不夠牛骨香，喜歡濃味的可再加多點份量。

Beef bone is the main ingredient here. The meat is not as flavorful as the bone. To have a richer taste, can add more beef bones in.

做法

1) 牛腰及骨先出水，再用水喉水清洗。雞也過一過滾水去血腥。
2) 洋葱、乾葱及薑片放入焗爐用 250℃ 上火焗 20 分鐘至焗香（如沒有焗爐可用白鑊炒香）。其他香料用白鑊中火烘幾分鐘至香味散出，備用。
3) 把出了水的肉和骨放入加了 4.5 公升水的湯煲，焗香了的蔬菜配料和所有香料放入煲魚袋內。加冰糖，煲要開蓋用小火煲 6 小時以上煮成湯底（時間愈長愈出味）。
4) 煮了兩個小時的牛腰可先拿出來，切片備用。
5) 煲好湯最後加入魚露及鹽、糖調味。
6) 把河粉白灼一會再放入湯碗，加上生牛肉及牛腰片、再倒入滾熱的湯底，鋪上洋葱條及芫荽葱粒。
7) 食時可隨意加入青檸汁、芽菜、金不換及指天椒粒。

「有記」麵家的越南河粉本人極力推薦，全港有多間分店。

I personally recommend the Vietnamese flat rice noodle of "Yau Kei" noodle house. It has many branches all over in Hong Kong.

Steps

1) Blanch the beef shank and beef bone. Rinse the beef shank and bone with running tap water after blanched. Quick blanch the chicken to reduce the taste of blood.

2) Pre-hit the oven with 250 °C with top grill, bake the onion, shallot and ginger for 20 minutes until golden and fragrant (if there is no oven, toast in dry wok until fragrant). Toast other ingredients in a dry wok with medium heat for a few minutes until fragrant. Set aside.

3) Put the beef shank and bone in soup pot and added with 4.5 liters water. Put the baked side ingredients and all spices in a food filter bag. Put into pot. Add the rock sugar. Uncover the pot and simmer for 6 hours or more to make the soup base (the longer the simmering the more flavorful the soup).

4) Dish up the beef shank first after 2 hours cooked in the soup. Slice and set aside.

5) Add the fish sauce, salt and sugar for seasoning when the soup done.

6) Blanch the rice noodles. Transfer to the soup bowl. Add the raw beef and cooked beef shank. Add boiling hot soup. Add the onion, diced coriander and spring onion. Serve.

7) Add some lime juice, bean sprout, sweet basil and diced fresh red chilli upon serving.

越南魚露用 40-50 度的最好。

The best Vietnamese fish sauce is around 40-50 °C.

榨青檸汁最好是皮向下肉向上，青檸汁經過皮再流入湯中最清香，檸檬有澀味不宜用這方法榨下。

The best way to squeeze the lime is the skin face down, let the juicy past by the skin than drop on the food will brings better favor.

越南 豬皮絲卷

Vietnamese Pig Skin Rice Rolls (Bì cuốn recipe)

我自己最喜歡的家鄉菜，可以在香港吃到的餐廳已經愈來愈少了，唯有在家自己做。越南材料可以到旺角煙廠街街市花姐越南雜貨店去買，檔口小小但應有盡有。

This is my favourite hometown dish. Fewer and fewer restaurants in Hong Kong are serving this. Best is to make it at home. Vietnamese ingredients are available at the store owned by sister Hua located in Mongkok Yin Chong street market. Though the store is small, it has everything.

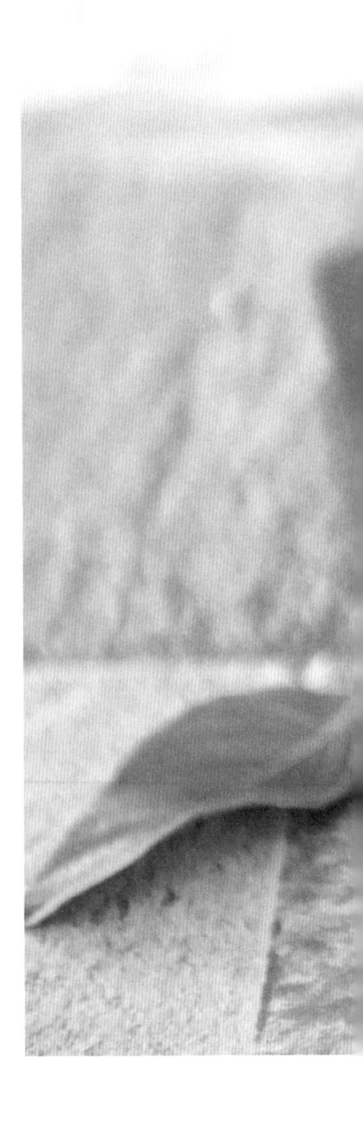

材料（2-4 人份量）

扎肉 / 白灼煮熟的豬肉 70 克（切幼條）
豬皮絲 70 克
炒碎米粉 3 湯匙
蒜頭 8 瓣（切蓉）
金不換 5 棵（只要葉）
唐生菜 1 棵（洗淨撕成片）
米紙 6 塊

Ingredients (Serves 2-4)

70g pork roll / poached pork (shredded)
70g shredded pig skin
3 tablespoons fried rice powder
8 cloves garlic (minced)
5 stalks sweet basil (use only the leaves)
1 stalk Chinese lettuce (rinse and tear to pieces)
6 Rice papers

Nước mắm 魚露材料

越南魚露 1 湯匙
糖 1 湯匙
水 3 湯匙
青檸汁 / 米醋 1 湯匙
指天椒 2 隻（切粒）
蒜頭 3 瓣（切蓉）
鹽少許

Fish sauce ingredients

1 tablespoon Vietnamese fish sauce
1 tablespoon sugar
3 tablespoons water
1 tablespoon lime juice / rice vinegar
2 fresh red chillis (diced)
3 cloves garlic (minced)
Pinch of salt

做法

1) 豬皮絲用食用水加少許鹽，洗淨瀝乾備用。
2) 切好的 2/3 蒜蓉用來炸做成熟蒜（炸蒜小心太大火易燶），其餘的留起作生蒜，備用。
3) 先做魚露，用 1 湯匙熱水把糖溶掉，多加 2 湯匙食用水、魚露、青檸汁 / 米醋攪勻，最後加入指天椒及生蒜蓉。調勻後放入冰箱雪凍備用。
4) 扎肉 / 豬肉絲加入豬皮絲內，再加入生、熟蒜，少許鹽、糖；一邊伴勻，一面逐少加入炒碎米粉。要令整個絲夠乾身便可。

皮卷要盡量包實一些，吃下去才有口感。

Tightly roll up rice roll to give it a better texture.

Steps

1) Soak and rinse the shredded pig skin with drinking water added with pinch of salt. Drain and set aside.
2) Deep-fry 2/3 of the minced garlic (do not use high heat as garlic gets burned easily). Set aside the rest of the raw garlic.
3) Making the fish sauce: dissolve the sugar in 1 tablespoon hot water. Add 2 tablespoons drinking water, fish sauce, lime juice / rice vinegar. Stir well. Add fresh red chillies and raw garlic. Stir well and chill in fridge.
4) Add the shredded port roll or pork to shredded pig skin in a big bowl. Add the raw and cooked garlic, pinch of salt and sugar. Toss well and add the fried rice powder slowly until the mixture become dry.

5) 開始捲時先張米紙浸入食用水幾秒令它輕微軟身，平放在碟上，鋪上略帶壓扁的生菜一塊，放上豬皮絲後再鋪上金不換。
6) 先從底向上摺一下，再左、右兩邊向內摺入，然後由底一直向上捲即成。食時可切開一半，用手拿起沾一口魚露吃一口便可。

5) Soak the whole sheet of rice paper in drinking water for few seconds until the rice paper has become a bit soft. Lay flat on a plate. Top with lightly flattened Chinese lettuce. Add the shredded pig skin mixture. Add the basil.
6) Fold the rice paper once from the bottom. Fold from left and right to the center. Roll up from the bottom. Upon serving, cut rice roll into half. Dip rice roll in fish sauce with each bite.

Asian Gourmet

亞洲風味

米紙蝦卷

Vietnamese Prawn Rice Rolls

一個非常簡單的前菜，食材亦比較容易買到，我的外國朋友特別喜歡。
This is a very simple appetizer and the ingredients are relatively easy to get. My foreign friends are especially fond of it.

材料（2-4 人份量） / Ingredients (Serves 2-4)

白灼煮熟的蝦 70 克
泰國粉絲 70 克（浸軟）
炸蒜 3 湯匙
金不換 5 棵（只要葉）
薄荷葉 3 棵
青瓜 1 條（切幼條）
唐生菜 1 棵（洗淨撕成片）
米紙 6 塊

70g poached prawn
70g Thai vermicelli (soaked in water until tender)
3 tablespoons fried garlic
5 stalks sweet basil (use only the leaves)
3 stalks peppermint leaf
1 cucumber (shredded)
1 stalk Chinese lettuce (rinse and tear into pieces)
6 Rice papers

Nước mắm 魚露材料 / Ingredients for fish sauce

越南魚露 1 湯匙
糖 1 湯匙
水 3 湯匙
青檸汁 / 米醋 1 湯匙
指天椒 2 隻（切碎細粒）
蒜頭 3 瓣（切蓉）

1 tablespoon Vietnamese fish sauce
1 tablespoon sugar
3 tablespoons water
1 tablespoon lime juice / rice vinegar
2 fresh red chillis (diced)
3 cloves garlic (minced)

做法

1) 切好的蒜蓉 2/3 用來炸做成熟蒜（炸蒜小心太大火易燶），其餘的留起作生蒜，備用。泰國粉絲用熱水快速灼煮 30 秒，拿起用食用水過冷河後瀝乾，備用。
2) 先做魚露，用 1 湯匙熱水把糖溶掉，多加 2 湯匙食用水、魚露、青檸汁 / 米醋攪勻，最後加入指天椒及生蒜蓉。調勻後放入冰箱雪凍。
3) 開始捲時先張米紙浸入食用水幾秒令它軟身，平放在碟上，鋪上略帶壓扁的生菜，再放 3 隻蝦、金不換、青瓜、薄荷葉、粉絲及蒜。
4) 先從底向上摺一下，再左、右兩邊向內摺入，然後由底一直向上捲即成。食時可切開一半，用手拿起沾一口魚露吃一口便可。

Steps

1) Deep-fry 2/3 of the minced garlic (pay close attention not to use high heat to avoid garlic from getting charred). Set aside the rest of raw garlic. Blanch the Thai vermicelli for 30 seconds. Dish up vermicelli and soak in cold water, drain.
2) Making the fish sauce: dissolve the sugar in 1 tablespoon hot water, combine with 2 tablespoons drinking water, fish sauce, lime juice / rice vinegar. Add fresh red chilli and raw garlic. Mix well and chill in fridge. Set aside.
3) Soak the rice paper in drinking water for a few seconds to soften it. Lay flat on a plate. Top with lightly flattened Chinese lettuce. Add 3 prawns, basil, cucumber, mint leaf, vermicelli and fried garlic.
4) Fold the rice papaert once from the bottom to cover the ingredients. Fold from left and right to the center. Roll up from the bottom. Cut into half upon serving. Dip in fish sauce before each bite.

配料方面很隨意，可用煮熟的蝦、牛、煎香的豬頸肉、再加青瓜、芽菜、煮熟的粉絲等隨意配搭……

Feel free to combine various ingredients such as cooked prawns, beef or pan-fried sliced pork neck with cucumber, bean sprout and cooked vermicelli to make the filling.

Asian Gourmet
亞洲風味

韓式 燜辣牛肋條

Spicy Braised Beef Rib Strip in Korean Style

我自己近這十年十分嗜辣，完全不懂吃辣的我由兩滴辣椒油開始到現在可以吃半茶匙。個人認為一點的辣的確可以提升了某些菜式的味道，但太辣卻做會完全蓋過了。所以我喜歡的都是一些微辣的食譜，以下這個亦是香微辣為主，不想吃辣亦可以不放辣椒醬及粉去做，而且很好下飯。

The past 10 years have seen me turning from someone who could just bear with two drops of chilli oil to the one who can enjoy as much as 1/2 teaspoon at each serving. It is my personal opinion that some spiciness can definitely enhance the taste of various dishes, but not to the extent that it becomes overwhelming. As such I have always preferred recipes that are mildly spicy. This is one such recipe that is aromatic and not too spicy. For a non-spicy version, omit the chilli sauce and powder altogether. It goes really well with rice.

材料（4 人份量）

牛肋條 10 條（約 600 克 切塊）
甘筍 2 條（切塊）
白蘿蔔半條（切塊）
金菇菜 1 包
洋葱 1 個（切條）
蒜頭 3 瓣（切片）

Ingredients (Serves 4)

10 strips beef rib strips (about 600g, cut into pieces)
2 carrots (cut into pieces)
1/2 radish (cut into pieces)
1 pack enoki mushroom
1 onion (shredded)
3 cloves garlic (sliced)

調味料

韓國辣椒醬 1.5 － 2 湯匙
（視乎燜完後湯汁有多少）
韓國辣椒粉 2 茶匙
（隨個人嗜辣程度去加減）
味醂 2 湯匙
牛高湯 3.5 杯
（或可用 1 粒牛精加 3.5 杯水代替）
鹽少許

Seasonings

1.5-2 tablespoons Korean chilli sauce
(depending the amount of leftover sauce when braising is done)
2 teaspoons Korean chilli powder
(adjust the portion according to personal preference)
2 tablespoons mirin
3.5 cups beef broth
(can be replaced by 1 cube beef stock added with 3.5 cups water)
Pinch of salt

做法

1) 牛肋條先出水，備用。下油，先炒香洋葱及蒜片。再加入甘筍及白蘿蔔炒香；再放入出了水的牛肋條略炒勻。
2) 加入牛高湯，味醂，水滾後轉小火。小火蓋上蓋料燜 1 小時
3) 之後再加入辣椒醬及辣椒粉，攪勻。再蓋上蓋多燜 15 分鐘至牛肉腍身便可。最後加入金菇菜多煮 5 分鐘。試味，不夠鹹才加少許鹽調味，即成。

1.1

1.2

3

Steps

1) Blanch the beef rib strips. Set aside. Heat a little oil. Stir-fry the onion and garlic until fragrant. Add the carrot and radish. Stir-fry until fragrant. Add the beef rib. Stir-fry a bit.
2) Add the beef broth and mirin. Bring to boil. Simmer and braise for 1 hour with lid.
3) Add the chilli sauce and chilli powder. Stir well. Cover the pot and braise more 15 minutes until the beef is tender. Add the enoki mushroom. Cook for 5 minutes more. Add a little salt if necessary. Serve.

간장게장
韓國醬油蟹

Korean Raw Crab Marinated in Soy Sauce

在韓國吃過真是回味無窮，韓國朋友的媽媽也傳授了這食譜給我。但生的蟹真是有點難度。有天經過街市發現海藍蟹其實也是類似品種，便膽粗粗地買來做。味道的確是一樣，但蟹膏不夠韓國的多。韓國因天氣比較冷所以長期都「爆膏」的海藍蟹，但香港便要入秋後了。

When I tasted this dish in Korea, it was truly unforgettable. My Korean friend's mother even taught me the recipe. But to be honest, raw crab is quite difficult to buy to make this dish. One day when I was in the wet market, I discovered that blue swimming crab is quite similar to the Korean crab specie. So I decided to experiment with it. Though the taste was the same, the crab roe was not as much as that one I tried in Korean. Due to the colder weather, Korean blue crab is full of crab roe whole year round. But Hong Kong's will have to be from autumn onwards.

材料（2-4 人份量）

活海藍蟹乸 2 隻
洋葱 2 個（切塊）
大紅椒 3 條 （切塊）
蒜頭 2 個 （原瓣拍扁）
蘋果 1 個（去皮切塊）

Ingredients (Serves 2-4)

2 live female blue sea crabs
2 onions (cut into pieces)
3 big red chillis (cut into pieces)
2 bulbs garlic (smashed)
1 apple (peeled and cut into pieces)

調味料

水 500 毫升
韓國豉油 350 毫升
韓國味醂 150 毫升
韓國粟米糖漿 250 毫升
韓國白酒 250 毫升

Seasonings

500ml water
350ml Korean soy sauce
150ml Korean mirin
250ml Korean corn syrup
250ml Korean white wine

做法

1) 蟹先用水刷洗乾淨，放入冰箱雪 30 分鐘把蟹凍暈。
2) 把所有調味料倒入煲，開火，再加入蘋果、一半的洋葱、蒜粒及辣椒。煮滾後轉中慢火煮半小時至味道滲出後熄火後乘涼。乘涼後的醬油隔渣備用。
3) 蟹要肚向上放入盒，再把涼了的醬油倒入浸過蟹身。

Steps

1) Scrub and rinse the crabs. Chill in the freezer for 30 minutes to render the crabs inactive.
2) Pour all the seasoning ingredients into a pot. Turn on the heat. Add the apple, half of the onion, garlic and chilli. Bring to boil. Adjust to medium low heat and simmer for half an hour until flavors are released. Turn heat off. Let cool. Strain and set aside the soy sauce.
3) Place the crabs stomach side up in a container. Add the cooled soy sauce to cover the crabs.

醬汁要每天煮滾才可以殺菌。

The sauce has to be boiled daily to kill all bacteria.

4) 把餘下一半的蒜粒、洋葱及辣椒加入。入雪櫃先浸一日。
5) 浸過一日的醬油蟹，先把醬油倒出鍋中，再滾熱後乘涼，再倒回蟹中再放入雪櫃多浸一日。每天重複以上步驟至蟹共浸了 3 天便可。
6) 食時先去厴，再開蓋，去鰓。再剪開鉗及把身體一開四即可享用。

4) Add the remaining garlic, onion and chilli. Chill in fridge for a day.
5) Pour the soy sauce into a pot. Bring to boil and let cool. Pour over the crab. Chill the soy sauce crab in fridge for another day. Repeat these steps to complete 3 full days of soaking.
6) Upon serving, remove the innards, lift up the shell, remove the lungs, cut the claws and quarter the crab body. Serve.

不想做蟹可以用蝦、鮑魚、雞蛋代替，鮑魚及蛋請先煮熟才浸。這些食材只需浸天一天便很入味。

Feel free to replace crab with prawns, abalone or egg. Cook abalone and eggs before soaking in the soy sauce. These ingredients take only 1 day of soaking to fully absorb the flavors.

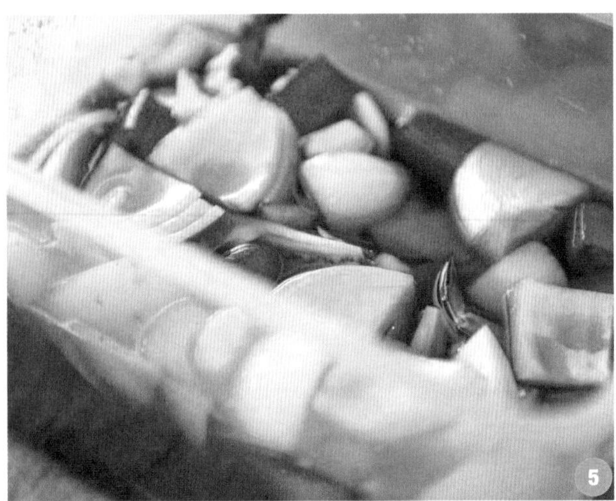

Asian Gourmet

亞洲風味

星洲蟹米線

Crab with Rice Noodles in Singapore Style (Crab Bee Hoon)

去新加坡必食蟹吧！而這道菜式亦是我一定會吃的。奶白的蟹湯加入米線，味道十分之協調。想也想不到這個菜式是十分容易煮，宴客又想簡單有排場，這是其中一個很好的食譜。

Crab is a must-eat when once visits Singapore. And this dish is also something I will definitely eat. Rice noodle is added to the milky white crab soup. The two ingredients blend really well together. It is surprisingly easy to make, making it a good choice if one wishes to have some simple yet guest-worthy dishes.

材料（2-4 人份量）

肉蟹1隻（約1斤半以上）
米線 1 斤（約 600 克）
菜心少許（隨意）
雞湯 800 毫升
淡奶 180 毫升
薑片 5 片
蒜頭 4 瓣（拍扁）
牛油 2 湯匙

Ingredients (Serves 2-4)

1 mud crab (more than 1.5 catties)
1 catty (600g) rice noodle
Some Choy Sum
(according to personal preference)
800ml chicken broth
180ml evaporated milk
5 slices ginger
4 cloves garlic (smashed)
2 tablespoons butter

調味料

紹興酒 2 湯匙
白胡椒粒 1 湯匙（壓碎）
魚露 1 湯匙
糖 1 茶匙

Seasonings

2 tablespoons Shaoxing wine
1 tablespoon white peppercorn (crushed)
1 tablespoon fish sauce
1 teaspoon sugar

煮蟹一定要用大火，否則肉質會霉。

To prevent the crab meat from turning mushy, it is a must to cook crab with high heat.

做法

1) 肉蟹煮前斬件洗淨，備用。
2) 下油，炒香薑片及蒜粒，再加入蟹件轉大火炒均。
3) 圍邊加入紹興酒，略炒至酒揮發。
4) 加入雞湯、牛油、白胡椒粒、糖、魚露，攪勻後蓋上蓋煮 3 分鐘至蟹八成熟。
5) 再加入菜心多煮 2 分鐘。最後放入米線多煮 1 分鐘便可。
6) 最後放入淡奶，即成。

蟹會在死後半小時便開始霉，請在煮前才劏蟹才能保持肉質厚實。

The crab meat will begin to turn mushy when the crab is dead for more than 1/2 hour. To preserve the firm meat texture, kill the crab just before cooking.

Steps

1) Chop and rinse the crab. Set aside.
2) Heat oil. Stir-fry the ginger and garlic until fragrant. Add crab pieces. Stir-fry over high heat.
3) Drizzle wine along edge of wok. Stir-fry until alcohol evaporates.
4) Add the chicken stock, butter, white peppercorn, sugar, fish sauce. Stir well. Cover wok. Cook for 3 minutes until crab is almost done.
5) Add the Choy Sum. Cook for 2 minutes. Add the rice noodle and cook for 1 minute.
6) Add the evaporated milk. Serve.

Asian Gourmet

亞洲風味

蚵仔麵線

Taiwanese Oyster Thin Noodles

我到台北，必定去食碗台灣麵線。慶幸朋友找到一家買手工麵線的店，我現在在香港也可以做。台灣人他們是用柴魚粉（即是鰹魚粉）做湯底、再加肉精、蝦精、香菇粉……我自己則改良成這家庭簡單版。

Each visit to Taipei will see me enjoying a bowl of the renowned Taiwanese thin noodle. It was a blessing that my friend was able to find a hand-made thin noodle shop so I can now replicate this dish in Hong Kong! Taiwanese people use ground bonito to make the soup base before adding the meat seasoning, prawn seasoning and mushroom powder. This recipe is a simplified version suitable for home cook.

材料（2 人份量）

手工麵線 80 克
蠔仔 200 克
芫荽 3 棵（切碎）

醃料

粟粉 1 茶匙

調味料

大地魚粉 / 柴魚粉約 2 茶匙
蝦子 1 茶匙
香菇粉 1 茶匙
老抽 1 湯匙
蠔油 1 湯匙
糖 1 茶匙
黑醋 1 湯匙
麻油少許
水 1.2 公升
辣椒油少許

生粉水芡

生粉 2 湯匙
水 6 湯匙

Ingredients (Serves 2)

80g handmade noodle Mian Xian

200g baby oyster

3 stalks coriander (finely chopped)

Marinade

1 teaspoon corn starch

Seasonings

2 teaspoons ground dried bonito/ ground Dried plaice

1 teaspoon dried shrimp roe

1 teaspoon mushroom powder

1 tablespoon dark soy sauce

1 tablespoon oyster sauce

1 teaspoon sugar

1 tablespoon black vinegar

A little sesame oil

1.2 liters water

A little chilli oil

Potato starch solution

2 tablespoons potato starch

6 tablespoons water

可自由加入喜愛食材去取代蠔仔，如豬大腸，豬肚，豬紅，魚蛋等……當然要因應食材煮熟的時間去改變放入次序。

Feel free to replace baby oysters with other ingredients such as pig intestine, pig stomach, pig's blood cube or fish ball according to personal preference. Take note that the sequence of adding these ingredients to the soup base needs to be adjusted according to how much cooking time is needed for each ingredient.

做法

1) 蠔仔洗淨，用 1 茶匙粟粉醃洗一會，再沖水洗淨，備用。
2) 下水煮滾，放入大地魚粉、蝦子、香菇粉、黑醋、蠔油及糖。
3) 加入麵線煮 2 分鐘。
4) 最後放入蠔仔略煮 30 秒，加入老抽調色，最後用生粉水打芡。
5) 吃時加入芫荽、黑醋、麻油及辣椒油便可。

Steps

1) Rinse the baby oysters, marinate with 1 teaspoon corn starch for a while. Rinse well and set aside.
2) Bring water to boil. Add the ground bonito/ground plaice, shrimp roe, mushroom powder, black vinegar, oyster sauce and sugar.
3) Add the thin noodles. Cook for 2 minutes.
4) Add the baby oysters and cook for 30 seconds. Add the dark soy sauce. Stir-in potato starch solution to thicken the soup.
5) Add the coriander, black vinegar, sesame oil and chilli oil upon serving. Serve.

2
3
4.1
4.2

Specialty Soups
特色湯類

COOKTOWN

花膠濃雞湯

Fish Maw with Chicken Milky Soup

前幾年很少地方做這個濃雞湯，直至好友那大熱的火鍋店用這做湯底後，即時又一窩蜂地抄襲到成行成市。這個湯底，的確是十分美味，做法有點繁複，但出來的效果卻十分值得一試。

Most of the restaurants do not serve this milky chicken soup for a while. But since my good friend's popular hot pot restaurant started using this broth as the soup base. All of a sudden it is served everywhere. This soup is truly delicious. Though the steps are quite painstaking, the outcome is well worth the effort.

材料（2-4 人份量） / Ingredients (Serves 2-4)

材料	Ingredients
瘦肉 500 克	500g lean pork
金華火腿 3 片	3 slices Jinhua hams
雞 1.5 隻	1.5 chickens
花膠 3 塊	3 pieces fish maw
水 3.5 公升	3.5 liters water

調味料 / Seasonings

調味料	Seasonings
鹽 2 茶匙	2 teaspoons salt

浸發花膠方法可參照 p.105「正確發花膠方法」。

Refer to p.105 for the method of rehydrate the fish maw.

本人建議用鱈魚膠煲湯，價錢便宜又比較不易煮溶。

I recommend the use of cod fish maw. The price is reasonable and it does not dissolve too quickly during the cooking.

做法

1) 先把瘦肉、雞洗淨。將瘦肉凍水下，汆水。雞用熱水過一過去血水便可。
2) 煲水，再落瘦肉及全部雞；用慢火煲 1 小時。
3) 取出 1 隻雞再另找鑊下油中大火把雞煎炒至金黃，再灒煲內滾熱的雞湯及金華火腿於鑊中，中大火滾 20 分鐘至奶白色。
4) 中大火轉小火滾 15 分鐘至湯奶白色，放入花膠及另外之前煲湯剩下的半隻雞多煮 10 分鐘，即成。

2

3.1

Steps

1) Rinse the lean pork and chickens. Blanch the lean pork and quick blanch the chickens to remove the blood.
2) Add the water to pot. Add the blanched lean pork and chickens and Jinhua hams. Simmer for 1 hour.
3) Dish up 1 whole chicken from the soup. Pan-fry in a wok until golden. Pour the boiling hot chicken broth and the hams into the wok and cook over high heat for 20 minutes until soup turns milky.
4) Adjust to low heat for 15 minutes until soup turns milky. Add the fish maw and the rest half chicken. Adjust to low heat and simmer for 10mins more. Add seasoning. Serve.

3.2

杏汁豬肺湯

Chinese Apricot Kernel and Pig's Lung Soup

每個人也會覺得煮這個湯太複雜，的確要洗豬肺真的好麻煩，但現在很多豬肉檔也會代勞，方便很多了。

This is a soup that everyone will find too complicated to make. It's true that the cleaning of pig's lung is quite tedious. The good news is that many pork sellers are now offering this service.

材料（2-4 人份量）

豬肺 1 個
南北杏 80 克（約半飯碗，隔晚浸軟）
元貝仔 10 粒（浸軟）
豬腱 1 斤 600 克（切件汆水）
雞腳 10 隻（汆水）
白果 20 粒（去殼去衣）

Ingredients (Serves 2-4)

1 pig's lung

80g Chinese sweet and bitter apricot kernel(about half rice bowl, soak overnight in water)

10 small dried scallops (soak in water until tender)

1 catty / 600g pork shank (cut into pieces and blanch)

10 chicken feet (blanch)

20 ginkgo nuts (remove shell and pith)

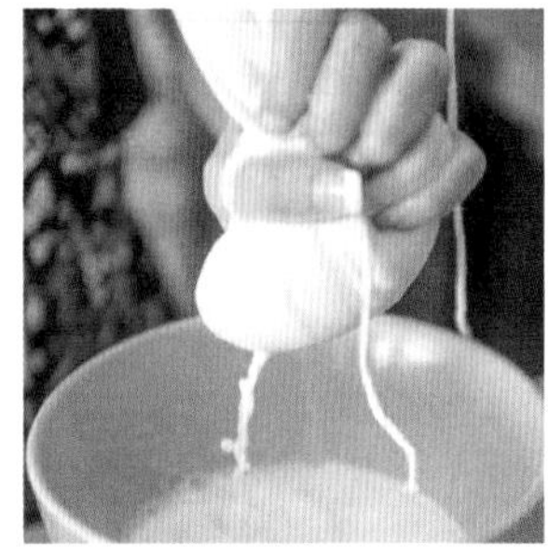

做法

1) 豬肺可以請豬肉檔代啤水至潔白。
2) 豬肺灌水再按壓出水份，重複多次，切大件，瀝乾水。
3) 豬肺件放入白鑊煎烘至微金黃色，盛起 。再出水，備用。
4) 南北杏隔晚浸軟，浸杏仁水保留。用 1:1.5 水份，將南北杏加水用攪伴機打成蓉、放入煲魚袋隔渣，取杏汁備用。
5) 除杏汁外，把所有材料加 3 公升水放入鍋煲 3 個小時。
6) 最後加入杏汁，多煲約 15 分鐘，加鹽調味即成。

Steps

1) Request the pork seller to repeatedly fill the pig's lung with water and squeeze dry until the pig's lung becomes white and clean.
2) Pipe water into the pig's lung and squeeze out the water. Repeat this many times. Cut thepig's lung into big pieces, drain.
3) Toast the pig's lung pieces in dry wok until light golden and blanch. Set aside.
4) Add the Chinese apricot kernel to the blender with the overnight soaked water. Add more water using the ratio of 1 part apricot kernel to 1.5 parts water. Puree the content. Transfer to the food filter bag. Strain and set aside the apricot kernel juice.
5) Put all the ingredients (except apricot kernel juice) and 3 liters water into pot. Simmer for 3 hours.
6) Add the apricot kernel juice. Simmer more 15 minutes. Add the salt to seasoning. Serve.

豬肺烘乾其間要按壓令其餘水份水滲出，請倒去水後再繼續烘、按壓至完全乾身、微金黃。

Press the pig's lung during the drying process to release the excess water. Drain and continue pressing and toasting until the pig's lung is completely dry and has turned light golden.

請保留浸元貝水去煲湯。

Retain the water used for soaking dried scallops to add to the soap.

Specialty Soups
特色湯類

鹹菜胡椒
豬肚湯

Pig Stomach, White Peppercorn and Pickled Mustard Soup

我爸爸是潮州人，每次去吃潮州菜都一定會叫這個，其實在家做真是「易過借火」。

My dad belongs to the Chiuchow clan. This is a must-order dish whenever he visits a Chiuchow restaurant. It is really easy to cook it at home.

材料（份量 2-4 人）

豬肚 1 個 （急凍便可）
鹹酸菜 300 克 （用水浸洗）
白胡椒粒 2 湯匙（壓碎）

Ingredients (Serves 2-4)

1 pig stomach (frozen one will do)
300g pickled mustard (soak and rinse)
2 tablespoons white peppercorn (crushed)

以上是潮式煮法，喜歡的可以多加腐竹和白果一起去煲。

This is a Chiuchow recipe. Feel free to add some dried tofu skin and gingko nut to the cooking.

怕辣可以減少白胡椒份量。

Reduce the white peppercorns for those who can't take spicy food.

做法

1) 急凍豬肚以生粉及鹽內外清洗乾淨後出水，備用。
2) 將豬肚放入鍋內汆水，直至水滾煮約 3 分鐘，撈起沖洗乾淨，備用。
3) 煲加入 2.5 公升清水，放入所有材料，小火煲 2 小時，最後加鹽調味便可。

Steps

1) Scrub the pig stomach inside and out with potato starch and salt. Rinse them well.
2) Put the pig stomach into a pot. Blanch and Dish up. Rinse.
3) Pour 2.5 liters water into a pot. Add all the ingredients. Simmer over low heat for 2 hours. Add the salt. Serve.

杜仲巴戟豬骨湯

Du Zhong, Ba Ji and Pork Bone Soup

我爸爸今年 86 歲，仍然身壯力健。與此湯水是有極大關係！爸爸說自己一直飲用此湯最少一個月 1 至 2 次，男女佳宜。我當年坐月時也有飲用。此湯有補腰補腎、疏通血管、清除瘀血、補充鈣質之效。就算不是坐月也可以一家飲用！希望大家也可以像我爸爸一樣健健康康。

At the age of 86, my dad's health is amazingly good. This soup is one of the factors. He consumes it once or twice a month. It is good for both men and women. When I was in post-partum confinement years ago, I used to drink this soup too. It can help strengthen lower back and kidney, unclog blood vessels, remove blood stasis and help replenish calcium, suitable for the consumption of the whole family. It is my sincere wish that everyone can be as healthy as my dad.

材料（2-4 人份量）

杜仲 2 錢
巴戟 2 錢
狗脊 2 錢
牛七 2 錢
豬骨 1.5 斤（斬件）
蜜棗 5-6 粒（開邊去核）

Ingredients (Serves 2-4)

2 qian Du Zhong
2 qian Ba Ji
2 qian Gou Ji
2 qian Niu Qi
1.5 catties pork bone (chopped to pieces)
5-6 candied dates (halved and remove seed)

做法

1) 豬骨出水，藥材略沖洗。將所有材料凍水放入湯煲，水滾後大火先煲 15 分鐘，再轉小火煲 4 小時，加鹽調味即成。

Steps

1) Place the knife diagonally between the top Blanch the pork bone. Rinse the herbs. Put all the ingredients to pot with cold water. Bring to boil and cook over high heat for 15 minutes. Adjust to low heat and simmer for 4 hours. Add salt. Serve.

醃篤鮮

Cured Meat and Soybean Sheet Knot Soup (Yanduxian)

材料（2-4 人份量）

鹹肉 400 克（切粗條）
五花腩 1 斤（切粗條）
百頁結 1 包（熱水浸軟）
冬筍 1.5 斤（去皮切角）
小棠菜 8 棵（開邊）

Ingredients (Serves 2-4)

400g cured meat (cut into thick strips)
1 catty pork belly (cut into thick strips)
1 pack soybean sheet knot (soak with hot water)
1.5 catties winter bamboo shoot (peeled and cut into wedges)
8 stalks Shanghai pakchoy (halved)

做法

1) 五花腩連鹹肉出水後， 和冬筍加入 3 公升水，大火先滾 20 分鐘，再轉小火煲 2 小時至味道香濃。
2) 放入百頁結和小棠菜，多煮 5 分鐘即成。

Steps

1) Blanch the pork belly and the cured meat. Add the pork belly, cured meat and bamboo shoot into 3 liters of water. Bring to boil over high heat and cook for 20 minutes. Adjust to low heat and simmer for 2 hours until the flavors have come together.
2) Add the soybean sheet knots and Shanghai pakchoy. Cook for 5 minutes. Serve.

Western Delicacy

西式風味

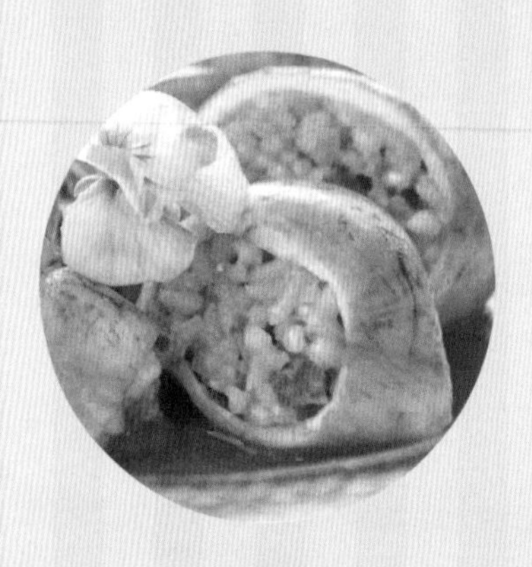

BROWN

黑松露 溫泉蛋意粉

Onsen Egg Spaghetti with Black Truffle

溫泉蛋與溏心蛋的形態是完全不同的，溏心蛋是蛋白實但蛋黃軟。溫泉蛋卻相反要蛋黃略凝固但蛋白軟軟的。然而蛋白與蛋黃均有不同的凝固點；蛋黃完全凝固的溫度是70℃，而蛋白卻要80℃才會完全凝固。而做溫泉蛋，蛋白輕微凝固點是62-65℃，而蛋黃則為65-70℃。所以要做到溫泉蛋效果，一般溫度是介乎63至67℃。而以63℃水浸45分鐘去慢煮為最理想效果。溫泉蛋也可放在保溫壺以68℃浸25分鐘。

Onsen(Hot sping) egg is totally different from soft-boiled egg. Half-boiled egg has firm egg white and soft egg yolk. Onsen egg is the reverse, the egg yolk is just set but the egg white is soft. Egg yolk fully sets around 70 °C and the egg white is at 80°C. For onsen egg, the egg white starts to set at 62-65 °C and the egg yolk at 65-70°C. Thus to create the onsen egg effect, the temperature is between 63 and 67 °C. To get the best result, soak the egg in the water with temperature of 63°C for 45 minutes. Onsen eggs can also be placed in a thermos pot for 25 minutes at 68°C.

材料（1 人份量）

雞蛋 1 隻
意粉 100 克
蒜頭 1 瓣（切片）
乾葱半粒（切碎）

Ingredients (Serves 1)

1 egg
100g spaghetti
1 clove garlic (sliced)
1/2 clove shallot (finely chopped)

調味料

黑松露醬 2-3 茶匙
黑松露油 1 茶匙
鹽 3 茶匙
雞粉 1 茶匙

Seasonings

2 or 3 teaspoons black truffle sauce
1 teaspoon black truffle oil
3 teaspoon salt
1 teaspoon chicken powder

如用蒸爐，你可以放蛋入蒸爐用 65℃ 蒸 45 分鐘去做溫泉蛋。

When using a steamer, put the egg into the steamer and steam at 65°C for 45 minutes to create an onsen egg.

做法

1) 雞蛋用 63℃水浸 45 分鐘做溫泉蛋。之後拿出。做好的溫泉蛋打在碗內，備用。
2) 煲水及下些油，3 茶匙鹽；煮意粉至八成熟後隔水，備用。
3) 鍋下油，炒香蒜片及乾葱，再加入意粉略炒，再加入雞粉，炒勻。
4) 再加入黑松露醬略炒。最後熄火，加入黑松露油，拌勻。
5) 上碟後把溫泉蛋放在中間，再在蛋面加些黑松露醬，即成。

Steps

1) Soak the egg in water with temperature of 63°C for 45 minutes to make an onsen egg. Dish up. Break the onsen egg in a bowl.
2) Add the water and a little oil in a pot. Add 3 teaspoons salt. Cook the spaghetti until 80% done. Drain.
3) Heat a little oil in wok. Stir-fry the garlic and shallot until fragrant. Add the spaghetti. Stir-fry briefly. Add the chicken powder. Stir well.
4) Add the black truffle sauce, stir-fry briefly. Turn heat off. Add the black truffle oil, stir well.
5) Transfer the spaghetti to serving plate. Place the onsen egg at the center. Add some black truffle sauce on egg. Serve.

Western Delicacy
西式風味

意大利大蜆意粉

Spaghetti Vongole

這是很最受歡迎的意大利菜式之一，食材便宜，既鮮味做法又簡單。但煮蜆要小心，若有一隻是壞的，便會把全個餸破壞了。中廚會喜歡用水略煮去沙去臭；但我自己在家煮喜歡保留蜆的原汁原味，所以唯有逐隻去聞有沒有異味了。

This is one of the most popular Italian dishes. The ingredients are not expensive, the taste is good and the steps are simple. Pay close attention while cooking the clams. A bad one will spoil the entire dish. Most Chinese chefs would blanch the clams in advance to release the sandy particles and bad smell. When making this dish at home, I prefer to retain the original flavor and juice of the clams. No other choice but to resort to smelling each clam to detect which are the bad ones.

材料（2 人份量）

大蜆 500 克（約 12 至 20 隻）
洋葱 1/4 個（切碎）
蒜頭 4 瓣（切碎）
意大利蕃荽 1 棵（切碎）
意粉 200 克

Ingredients (Serves 2)

500g big clam (about 12 to 20 pieces)
1/4 onion (finely chopped)
4 cloves garlic (finely chopped)
1 stalk parsley (finely chopped)
200g spaghetti

調味料

白酒 300 毫升
鹽 1/2 茶匙
糖 1 茶匙
辣椒粉半茶匙
（可隨口味加減或不放）

Seasonings

300ml white wine
1/2 teaspoon salt
1 teaspoon sugar
1/2 teaspoon chilli flake
(adjust the portion or totally omit, according to personal preference)

tips

意粉千萬不要過冷河，經凍水泡過的意粉外表會硬，煮意粉時醬汁不容易被吸入，意粉難入味。

Do not soak the spaghetti in cold water. Spaghetti that has been soaked in cold water has a hard exterior, renders it hard to absorb the sauce, the spaghetti will be less flavorful.

做法

1) 煲下水，加少許油及 1 茶匙鹽，滾起後下意粉，開蓋煮 8 分鐘，意粉約八成熟，隔水取出，備用。
2) 鍋中大火下油，先炒香蒜及洋葱，再下大蜆略炒。
3) 加入白酒，中火煮約 4 分鐘（視乎蜆大小），白酒汁收剩至一半。
4) 大蜆差不多半開口時加入鹽、糖、辣椒粉調味。
5) 馬上加入意粉把汁吸收（若汁太多可轉大火收汁），熄火再加少許蕃荽即成。

Steps

1) Add the water to a pot. Add some oil and 1 teaspoon of salt. Bring to boil. Add the spaghetti. Cook in uncovered pot for 8 minutes. Cook until spaghetti is 80% done. Drain and set aside.
2) Heat a pan and oil over medium high heat. Stir-fry the garlic and onion until fragrant. Add big clams and stir-fry briefly.
3) Add the white wine. Cook over medium heat with the lid cover for around 4 minutes (depending on size of clams). Cook until white wine is reduced by half.
4) When the clams are half opened, add the salt, sugar and chilli flake.
5) Add the spaghetti immediately to absorb the sauce and stir - fry (if there is too much sauce, cook over high heat to reduce it). Turn heat off and garnish with parsley. Serve.

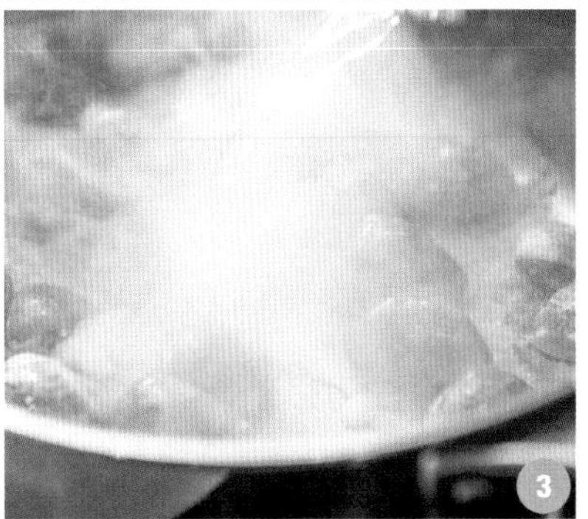

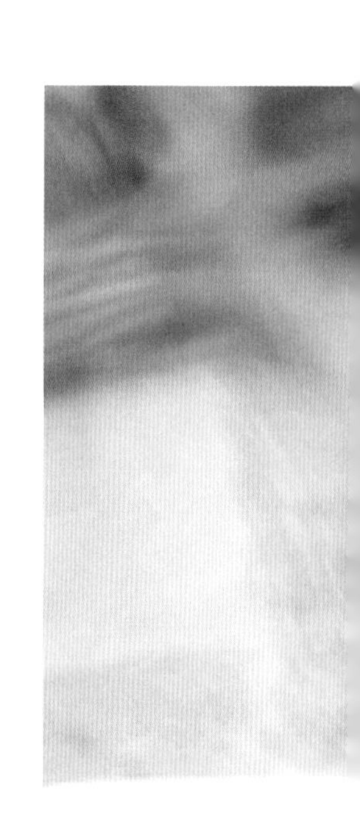

番茄龍蝦
手造意粉

Lobster with Home-made Spaghetti in Tomato Sauce

做菜時可以自己由零變出一整個菜式出來，滿足感是十分大。有空可以自製意粉，沒有的買現成也可。怕劏龍蝦的，買急凍亦可，最重要要選些連殼的，因為龍蝦殼是熬製意粉汁的主要材料。

It is very satisfying to create something from nothing. If there is time, make some spaghetti yourself. If not than just buy it. Not sure of how to dress the fresh lobster? Then get a frozen one. Just make sure to get the one with shell because lobster's shell is the main ingredient of the sauce for the spaghetti.

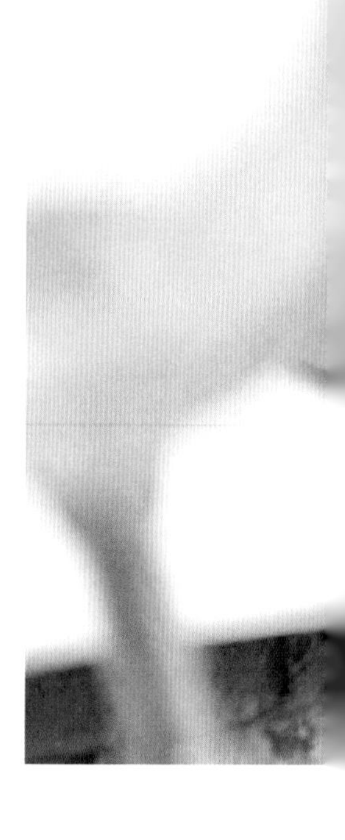

材料（1人份量）

波士頓龍蝦 1 隻
洋葱半個（切粒）
紅椒 半隻（切條）
甘筍 1 條（切粒）
西芹 1 段（切粒）
番茄粒 200 克
法國洋葱 2 粒（切碎）
蒜頭 5 瓣（切片）
蕃茜少許

調味料

清雞湯 400 毫升
月桂葉 2 片
百里香半茶匙
白酒 100 毫升
忌廉 20 毫升
糖 1 茶匙
鹽少許
黑胡椒少許
麵粉 1 湯匙
辣椒粉少許

意粉材料

00 麵粉 80 克
粗粒小麥粉 20 克
橄欖油 1 茶匙
鹽 少許
雞蛋 1 隻
凍水 2 茶匙（如需要）
牛油少許

Ingredients (Serves 1)

1 Boston lobster
1/2 onion (diced)
1/2 red bell pepper (shredded)
1 carrot (diced)
1 section celery (diced)
200g chopped tomato
2 french onions (finely chopped)
5 cloves garlic (sliced)
Pinch of parsley

Seasonings

400ml chicken broth
2 bay leaves
1/2 teaspoon thyme
100ml white wine
20ml whipping cream
1 teaspoon sugar
Pinch of salt
Pinch of ground black pepper
1 tablespoon flour
Pinch of chilli powder

Pasta ingredients

80g 00 flour
20g semolina flour
1 teaspoon olive oil
Pinch of salt
1 egg
2 teaspoons cold water (if needed)
A little butter

處理龍蝦

新鮮龍蝦先在頭下部劏一刀後，原隻放入滾水內蓋上蓋煮 2.5 分鐘，急凍的則只需解凍。之後拿出起肉去殼，肉切塊，殼及蝦頭切細留起。

To prepare the lobster

Make a cut at the head of the fresh lobster. Put whole fresh lobster in the boiling water. Cover the lid and cook for 2 .5 minutes. Only thawing is needed if frozen lobster is used. Separate meat from lobster shell. Cut the meat into pieces. Cut the shell and lobster head into small pieces.

手造意粉做法

1) 把 00 麵粉、粗粒小麥粉及鹽放在一起。
2) 橄欖油和雞蛋攪勻，若麵糰太乾可加入少許凍水。
3) 用保鮮紙蓋過麵糰，醒麵糰 30 分鐘。
4) 放入壓麵機先壓兩次。
5) 逐步壓扁至 6、7 度厚度便可，之後放入切麵機。
6) 可以掛起風乾，或可即刻煮。
7) 放入有油、鹽的水煮約 1.5 分鐘便可。

Making the spaghetti

1) Combine the 00 flour, semolina flour and salt.
2) Add the olive oil and egg. Mix well. Add a little bit water if the dough is too dry.
3) Rest the dough covers with a sheet of cling wrap for 30 minutes.
4) Put into the dough roller. Roll the dough twice.
5) Gradually roll the dough to thickness score of 6 or 7 will do. Transfer sheet dough to dough cutter to make the spaghetti.
6) The fresh spaghetti can be air-dried with a pasta drying rack or cooked immediately.
7) Put into boiling water added with oil and salt. Cook for 1.5 minutes for the fresh one.

龍蝦意粉做法

1) 下油，加入龍蝦殼炒勻，要盡量炒得香點乾身點，約需炒 4 分鐘。
2) 倒入白酒炒香。
3) 下洋葱、蒜、甘筍、西芹、紅椒、法國洋葱先炒香（請留起一點蒜片、法國洋葱、紅椒之後炒龍蝦肉用）。
4) 再加入雞湯、百里香、月桂葉及黑胡椒。
5) 先用大火煮 5 分鐘，再轉中小火煮約 10 分鐘至湯汁收至接近一半後隔渣，備用。
6) 鑊下牛油及鹽，炒香龍蝦肉至八成熟，拿起備用。
7) 鑊下牛油，下剩下的蒜片、法國洋葱、紅椒炒香。加入少許黑胡椒。
8) 加入番茄粒略炒，再加入之前的龍蝦湯 。加入辣椒粉、糖、鹽及 1 湯匙麵粉，最後下忌廉把汁煮至少許杰身。
9) 加入意粉及龍蝦肉。需要的話，可加些鹽作調味。回鑊炒勻放上碟後加些蕃茜，即成。

Steps

1) Heat a little oil. Add the lobster's shell pieces. Stir-fry until it is fragrant and a little dry, takes about 4 minutes.

2) Add the white wine. Stir-fry until fragrant.

3) Stir-fry the onion, garlic, carrot, celery, red bell pepper and french onions until fragrant (set aside some of the sliced garlic, french onions and red bell pepper for the stir-frying of lobster).

4) Add the chicken broth, thyme, bay leaves and black pepper.

5) Cook over high heat for 5 minutes. Adjust to medium low heat and cook for more 10 minutes until sauce reduces by half, strain the soup and set aside.

6) Heat the butter and salt in a pan. Stir-fry until lobster meat is 80% done. Set aside.

7) Add some more butter in the same pan. Add the remaining sliced garlic, french onions and red bell pepper. Stir-fry until fragrant. Add some black pepper.

8) Add the chopped tomato. Add the lobster broth in. Stir well. Add the chilli powder, sugar, salt and 1 tablespoon flour. Stir well with the cream to thickens the sauce.

9) Add the spaghetti. Cook for a while to the desired hardness. Add the lobster meat. Stir-fry briefly. Taste it. Add a little salt for seasoning if necessary. Transfer to plate. Add some parley on top. Serve.

6

8

9

香芒雞肉
酥皮卷

Mango and Chicken Puff Pastry Roll

這是我自創菜式，
做法容易，做 party food 前菜一流，
大人小朋友也會喜歡。

This recipe is my own creation.
It is an easy-to-make party appetizer that
both adults and kids will appreciate.

材料（2人份量）

雞胸肉 1 小塊（150 克）
酥皮 2 塊
芒果 100 克（切細粒）
小三色椒各 2 湯匙（切細粒）
西芹 1 條（切細粒）
碎巴馬臣芝士 10 克
忌廉 50 毫升
蛋黃 1 隻（攪勻）

調味料

紅椒粉半茶匙
糖少許
黑椒少許
鹽半茶匙

Ingredients (Serves 2)

150g chicken breast meat
(cut into small cubes)
2 pieces puff pastry
100g mango (cut into small cubes)
Bell pepper
(3 colours, each 2 tablespoons, cut into small cubes)
1 stick celery (cut into small cubes)
10g grated Parmesan cheese
50ml cream
1 egg yolk (beaten)

Seasonings

1/2 teaspoon paprika powder
Pinch of sugar
Pinch of ground black pepper
1/2 teaspoon salt

做法

1) 雞肉用鹽、黑椒及紅椒粉醃一會。
2) 中火以少許橄欖油起鑊，先炒香小三色椒及西芹粒，再下雞肉，接近熟時最後下芒果粒炒勻。
3) 拌入忌廉、碎芝士及少許糖把汁煮杰，再加黑椒，離火備用。
4) 焗盤上放上牛油紙，把現成酥皮放上。再於兩側約 1/3 闊度起每隔 1.5 cm 剘開至底部。
5) 中間放上炒好的蔬果雞粒連汁，把左邊第一條搭到右邊第三條的位置，跟着右面第一條又搭到左邊第三條，左右重疊織至底部成卷狀，最底那條要把底部橫向封密，撕一些多餘的把卷的頭也封好，塗上蛋漿。
6) 放進已預熱 180℃ 的焗爐焗 20 分鐘即成。

Steps

1) Marinate the chicken meat with the salt, ground black pepper and paprika powder. Let stand for a while.
2) Heat a little olive oil over medium heat in a pan. Stir-fry the bell pepper and celery until fragrant. Add the chicken breast meat. When it is almost done, add the mango. Stir-fry until even.
3) Add the cream, pinch of grated cheese and ground black pepper. Stir until the sauce thickens. Remove from heat.
4) Line baking pan with the parchment paper. Place the puff pastry on the paper. Cut 1.5 cm wide strips horizontally and leave the middle strip about 1/3 total width.
5) Place the vegetable chicken mixture along the centre strip, lift up the first pastry strip at the left and cross it to where the 3rd strip at the right is, lift up the first strip at the right and cross it to where the 3rd strip at the left is, repeat this until the base forms shape of a roll. Place the last strip horizontally to seal up the roll. Tear some excess pastry to cover up the head of the pastry roll. Brush on beaten egg yolk.
6) Put into oven preheated at 180 °C and bake for 20 minutes. Serve.

Western Delicacy

西式風味

鮮牛肉他他

Beef Tartare

HERMES

材料（1-2 人份量）

冰鮮瘦牛柳 200 克
蛋黃 1 隻
乾葱 2 湯匙（切蓉）
蕃荽 1.5 湯匙（切碎）
酸豆 1 茶匙（切蓉）

Ingredients (Serves 1-2)

200g chilled beef fillet
1 egg yolk
2 tablespoons shallot (minced)
1.5 tablespoons parsley (chopped)
1 teaspoon Capers (minced)

醬汁料

芥末 1 茶匙
松露油 2 湯匙
茄汁 1 湯匙
Tabasco 6-8 滴
青檸汁 2 茶匙
糖 1.5 茶匙
喼汁半茶匙
鹽 1 茶匙

Seasonings

1 teaspoon mustard
2 tablespoons truffle oil
1 tablespoon ketchup
6-8 drops Tabasco
2 teaspoons lime juice
1.5 teaspoons sugar
1/2 teaspoon worcestershire sauce
1 teaspoon salt

牛柳要選瘦的，選冰鮮的比較新鮮。

Use the beef fillet that is chilled. It is fresher than the frozen one.

油及蛋黃可以隨意多加一點，牛肉要靠它們去做出黏黏的效果。

Feel free to add more oil and egg yolk. These two ingredients are needed to create the sticky effect.

做法

1) 牛柳切條再切粒，再剁成肉醬，備用。
2) 大碗放入蛋黃，芥末及少許鹽用叉慢慢攪勻，再放入松露油拌勻至成醬。
3) 加入乾葱、酸豆蓉及蕃茜再拌。再加入茄汁、tabasco、青檸汁及糖攪勻；最後加入牛肉拌勻。
4) 把牛肉放入模內再壓平，加上伴碟，即成。

Steps

1) Cut the beef fillet into strips. Dice and mince it. Set aside.

2) Put the egg yolk, mustard and pinch of salt in a big bowl. Beat with a fork and then slowly incorporate the truffle oil and stir well to make the sauce like the mayonnaise.

3) Add the shallot, capers and parsley. Mix well. Add the ketchup, Tabasco, lime juice and sugar. Stir until even. Add the beef. Mix well.

4) Transfer the beef mixture into a mould. Flatten it. Add the garnishing. Serve.

做完後請即食用以保衛生。

This dish is to be consumed immediately once it is made.

煙燻玫瑰
黑椒吞拿魚

Smoked Tuna Rose with Black Pepper

無論賣相，味道，絕對一流。做法卻又十分容易。再加上煙燻，輕易便變成了米芝蓮餐廳菜式。當然，沒有煙燻機的最後不做煙燻也可。

This dish tops the list in terms of presentation and taste but the making is really easy. With the smoking feature it becomes a dish worthy to be served in a Michelin restaurant. The smoking step is optional for those with no smoking machine.

刺身吞拿魚最好有 3 x 6 cm 高和濶。

Desired size of the tuna fillet should be 3cm height x 6cm width.

材料（2－4 人份量）

刺身吞拿魚 1 條（約 500 克）
黑胡椒碎 5 湯匙
法國芥末 1 茶匙
麻油 1 湯匙

Ingredients (Serves 2-4)

1 strip tuna sashimi (500g)
5 tablespoons ground black pepper, crushed
1 teaspoon dijon mustard
1 tablespoon sesame oil

汁醬材料

鰹魚汁 1 湯匙
日本刺生豉油 1 湯匙
日本芥末 1 茶匙

Sauce ingredients

1 tablespoon bonito sauce
1 tablespoon Japanese soy sauce for sashimi
1 teaspoon wasabi

做法

1) 用廚房紙把吞拿魚肉抹乾，長長切出薄薄的 6-8 片用來包玫瑰花作伴碟，放入冰箱備用。
2) 把麻油加入法國芥末攪勻，再在魚肉上四面掃上攪勻的法國芥末。
3) 把黑胡椒碎鋪在平碟上，把魚肉整條印滿黑椒，不要有空位。
4) 平底鑊白鑊大火燒紅，放入魚每面煎約 45 秒，拿出切片放在入碟中。
5) 把醬汁材料全部攪勻，平均淋在魚上。再加煙燻焗烤 10 分鐘，即成。

玫瑰花伴碟做法 Steps for making the rose

Steps

1) Pat dry the tuna with kitchen paper. Cut out 6-8 long thin slices of tuna fillet to form shape of rose. Chill in the fridge.
2) Combine the sesame oil with the dijon mustard. Brush all sides of tuna with the mixture.
3) Spread crushed ground black pepper in a plate. Place the tuna on top and coat the entire piece of it, do not leave any empty space.
4) Heat a dry frying pan until red hot. Put in tuna and pan-fry each side for 45 seconds. Dish up and cut into slices. Transfer to a plate.
5) Combine the sauce ingredients for making the sauce. Drizzle evenly on tuna. Use the smoking machine to smoke the tuna for 10 minutes. Serve.

酥皮忌廉
菠菜焗魚柳

Puff Pastry Baked Fish Fillet with Spinach

我自己十分喜愛吃酥皮，
其實配合在餸菜上也可以很夾，
上碟時又夠大方得體。

I am a fan of puff pastry.
It can be served together with other dishes as part of the meal.
The plating presentation is attractive as well.

材料（2 人份量） Ingredients (Serves 2)

材料	Ingredients
龍脷柳 1 條（切件）	1 piece sole fillet (cut into pieces)
西式菠菜葉 200-300 克	200-300g spinach leaves
急凍酥皮 1-2 塊	1-2 pieces puff pastry
忌廉 20 毫升	20ml whipping cream

調味料 Seasonings

調味料	Seasonings
黑椒碎 1/2 茶匙	1/2 teaspoon ground black peppercorn (crushed)
鹽 1/2 茶匙	1/2 teaspoon salt
蛋黃 1 隻	1 beaten egg
蛋白 1/2 茶匙	1/2 teaspoon egg white

貼士 tips

現成酥皮要焗 20 分鐘以上才會鬆起及金黃，所以魚柳不需煎熟。

The puff pastry takes 20 minutes of baking to turn golden and fluffy. As such the fish fillet does not need to be fully cooked during the pan-frying.

做法

1) 把龍脷柳洗淨抹乾水份，焗爐預熱180℃。
2) 預備菠菜，只要葉，洗淨用乾布吸乾水份備用。
3) 龍脷柳大火煎香兩邊至金黃色，加黑椒及鹽調味即可，不用煎熟。
4) 另一鍋中大火下油，加入菠菜，炒熟，加少許忌廉並調味，隔去水份，待用。
5) 將龍脷柳放於酥皮上，加菠菜於龍脷柳上，兩邊預留約1吋酥皮圍邊塗上蛋醬，再鋪上另一塊酥皮封好。最後在酥皮面再塗蛋醬，焗約20分鐘或至呈金黃色即可。

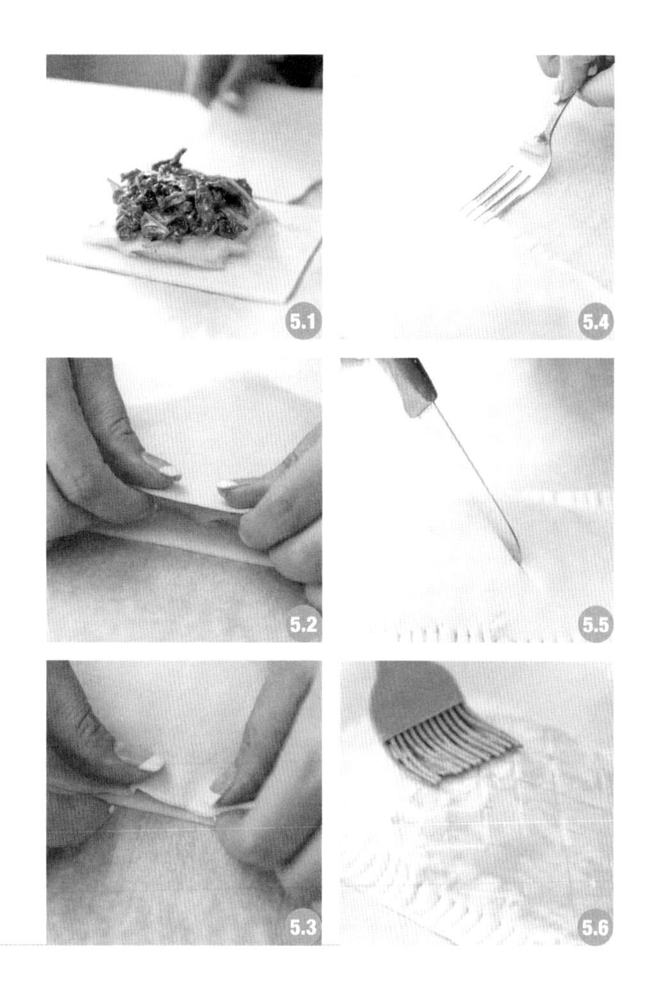

Steps

1) Rinse and pat dry the sole fillet. Preheat oven at 180 °C.
2) Use only the leaves of the spinach. Rinse well and pat dry with dry cloth.
3) Pan-fry the sole fillet until both sides turn golden. Seasoning with the ground black peppercorn and salt. Does not need to be fully cooked.
4) Heat oil in another pan over medium high heat. Add the spinach. Stir-fry until done. Season with a little whipping cream. Drain and set aside.
5) Place the fish fillet on the puff pastry. Top with the spinach leaves. Leave 1-inch of space at both sides of the pastry. Brush on egg wash. Cover with another piece of puff pastry. Seal the sides and brush the egg wash on top of the puff pastry. Bake in oven for 20 minutes or until golden. Serve.

西班牙
魷魚焗飯

Baked Stuffed Squid with Valencia Rice

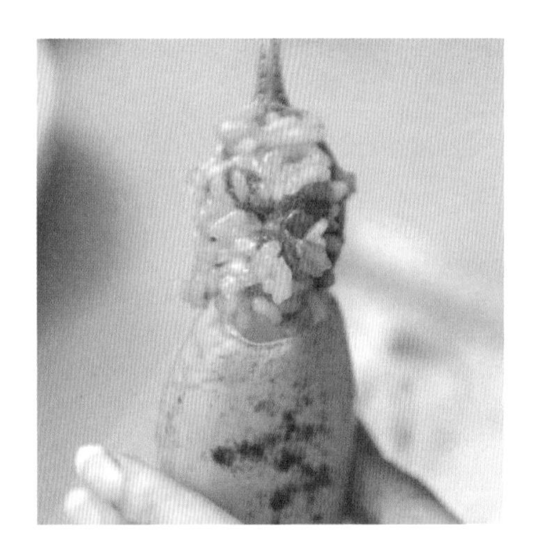

又是一個自創菜式，擺盤可以很漂亮，
海鮮控們一定要試試。

This is another of my creation
with a beautiful presentation,
a must-try for seafood lovers!

材料（2人份量）

魷魚2大隻
西班牙辣肉腸100克
番茄1個（切小粒）
洋葱半個（切碎）
蒜3瓣（切碎）
西班牙米200克
白酒50毫升
雞湯150毫升
蕃茜少許

Ingredients (Serves 2)

2 big squids
100g chorizo, diced
1 tomato, diced
1/2 onion, finely chopped
3 cloves garlic, finely chopped
200g Valencia rice
50ml white wine
150ml chicken broth
A little parsley, finely chopped

調味料

西班牙辣椒粉2茶匙
糖1茶匙
鹽1/2茶匙
辣椒片1茶匙
豉油2茶匙

Seasonings

2 teaspoons paprika
1 teaspoon sugar
1/2 teaspoon salt
1 teaspoon chilli flakes
2 teaspoons soy sauce

紅蝦汁材料

蝦頭12隻
白酒150毫升
干邑1湯匙
茄膏1.5湯匙
糖2茶匙

Prawn sauce ingredients

12 prawn heads
150ml white wine
1 tablespoon cognac
1.5 tablespoons tomato paste
2 teaspoons sugar

做法

1) 魷魚洗好，身留起成筒形。把魷魚鬚及翼切小粒。
2) 中大火下辣肉腸炒至出油，再加入蒜及洋葱炒香，再加番茄、魷魚粒略炒。
3) 轉小火加入米炒匀，再加入 50 毫升白酒略煮 1 分鐘，再加入雞湯，放入少許鹽、糖、1 茶匙西班牙辣椒粉炒匀。蓋上蓋煮 20 分鐘。
4) 魷魚身上塗匀豉油及灑上 1 茶匙西班牙辣椒粉，把煮好的飯再塞進魷魚筒入填滿壓實，再用牙籤封口。
5) 預熱 200℃上火的焗爐 15 分鐘，焗盤放一蒸架再放入魷魚，焗 20 分鐘，期間反一下。
6) 另一小鍋中大火下油，放蝦頭炒香後下干邑略炒，再下白酒炒匀後由它收汁至小於一半，最後加入茄膏及糖調味，汁有點杰便可。
7) 焗好的魷魚飯拿出來斜切厚片，放到碟上，再淋上紅蝦汁，即成。

Steps

1) Rinse the squids. Leave the body tube intact. Dice the squid tentacles and wings.
2) Stir-fry the chorizo over medium high heat to release the oil. Add the garlic and onion. Stir-fry until fragrant. Add the tomato and diced squid. Stir-fry briefly.
3) Adjust to low heat. Add the rice and stir-fry until even. Add 50ml white wine. Cook for 1 minute. Add the chicken broth. Add pinch of salt, sugar, 1 teaspoon paprika. Stir-fry until even. Cover the pan with lid and cook for 20 minutes.
4) Spread the soy sauce on squid body tube. Sprinkle with 1 teaspoon paprika. Firmly stuff squid body tube with rice. Seal the openings with toothpick.
5) Preheat the oven using the grill function at 200°C for 15 minutes. Place a steaming rack on baking tray. Put the squid stuffed on the rack. Bake for 20 minutes. Turn the squid body over half way through the baking.
6) Heat a little oil in a sauce pan. Stir-fry prawn heads until fragrant. Add the cognac. Stir-fry briefly. Add the white wine. Stir well until sauce reduces by more than half. Add the tomato paste and sugar as seasoning. Cook until sauce thickens.
7) Slice the squid stuffed into thick rings. Arrange on the plate. Drizzle with the prawn sauce. Serve.

Bonnie 入廚小貼士

1 炸物技巧

炸油溫度應在 140 至 180℃之間，炸海鮮可以在 150℃左右，炸肉則大概 180℃。若見油出煙表示過熱，這時下食材會很容易焦。剛下食材及炸好拿起前要用大火把油迫出。

Deep-Frying skills

The temperature of the deep-frying oil should be between 140 to 180 °C. For seafood can be around 150 °C and for meat around 180 °C. Wisp of smoke indicates overheating. Ingredients entering the hot oil at this point will be easily burnt. Use high heat just before the ingredients are removed from the hot oil to boil out the excess oil.

2 意粉應否過冷河？

千萬不要。意粉若過了冷河，麵條外層因冷卻會被封鎖，意粉汁醬便很難滲進，吃時會有「汁還汁、麵還麵」的感覺。麵條過冷河的作用是想保持彈口，所以用在湯麵中還可。若意粉那些只靠少量汁醬的是不太適宜。

Should spaghetti be soaked in cold water after the blanching?

Do not do so. Soaking blanched spaghetti in cold water will seal the outer layer of pasta, preventing the sauce absorb into the noodles. The noodles and sauce will not be well assimilated. The purpose of soaking cooked pasta in cold water is to preserve the al dente mouth-feel. It works for soup noodles but not applicable to spaghetti with little sauce.

3 如何處理海鮮食材？

很多人也以為買蜆回家後，要浸水去沙，這是不可行的。加些鹽亦沒有令蜆開口吐沙，只會令其因窒息至死而開口。若你浸幾小時後才煮分分鐘令蜆早死而變臭。所有鹹水海鮮，買回家後立即放入冰箱，到食前才用食水洗再即刻煮。用淡水洗過的海鮮不可存留太久。蜆可在吃前 20 分鐘才洗及浸水，令其剛開始窒息吐沙又未死去時立即烹煮為最佳。

How to handle seafood?

Most people have the notion that live clams should be soaked in the water to remove the sandy particles. Adding some salt will also not cause the clams to open the shells to release the trapped sand. It will only cause the clams to die of suffocation first when opening up the shells. If the clams are soaked for a few hours and then cooked, it would only cause the clams to die much earlier and smell bad. All saltwater seafood should be refrigerated immediately after being brought home, rinsed with drinking water just before cooking. Seafood that are rinsed with fresh water cannot be kept for too long. Clams should be rinsed and soaked in water 20 minutes before cook. The timing is just right when they have just started to suffocate and begin to release the trapped sand, yet not completely dead yet.

4 煮食材火喉控制

煮蟹及龍蝦一定要大火，否則肉會霉。煮雞、鮑魚一定要慢火，否則會過老及硬。

Heat control

The crab and lobster have to be cooked with high heat so that the meat will not become mushy. The chicken and abalone have to be cooked with low heat to prevent the meat from getting tough when overcooked.

5 煲湯技巧

除了西洋菜，所有食材應凍水落煲一起開火，才會令食材更易出味。煲老火湯在最後的半小時開中火大滾一下味道更出。若要煲出清澈的湯底如燉湯般，則要全程最小火，開蓋煲。湯水滾動愈大愈混濁。

Soup-cooking skills

Except watercress, all ingredients should be added into cold water before or just turning on the heat. This will allow the ingredients to better release the flavours. When making long-simmering soup with rich flavour, adjust to medium heat at the first half hour to bring the soup to vigorous boil. For a clear soup base similar to that of double-boiled soup, use low heat the entire cooking process without covering the lid. The more vigorous the boiling the more turbid the soup.

6 涼瓜下鹽

煮涼瓜時，怕太苦又不想出水失去口感的話，可用鹽把切好的涼瓜件醃幾下，再沖水後煮便可。

Adding salt to bitter gourd

When cooking the bitter gourd, to lessen the bitterness, instead of blanching the gourd that might affect the mouth-feel, briefly marinate sliced bitter gourd with salt and rinse well before cooking.

7 打芡

打芡的技巧

芡汁，是生粉與水的結合，可算是粵菜常用的步驟。看似簡單，但其實亦有技巧。

在粵菜中，勾芡一般稱為打芡，而打芡時可用的粉包括生粉、粟粉、藕粉、麵粉和馬蹄粉等……每一種芡粉有不同的特點，比如麵粉的芡會較脆身、馬蹄粉打的芡放涼後會爽脆且不會太平滑。而一般粵菜餐廳會用生粉芡，因為比較「杰身」，用其做炸物也比較脆口。家庭烹調則多用粟粉，醃肉打芡。而我自己打芡會用生粉，醃肉則用粟粉。

粉及水份量要視乎芡汁的濃稠度而定，我自己一般粉跟水的打芡比例是 1:3。而做炸漿是 1:1。

Adding thickener

Method

Thickener is a combination of potato starch with water and is commonly used in Cantonese cuisine. Though it looks simple, certain skills are involved.

These are various agents used to make thickener: potato starch, corn starch, lotus root powder, flour and water chestnut flour etc. Each has its own uniqueness. Thickener made with flour will create a crispier texture, thickener made with water chestnut flour is crunchier but not as smooth. Potato starch thickener is often used in Cantonese restaurants due to its stickier texture and the deep-fried stuff coated with this thickener has a crunchier mouth-feel. Home-cooked dishes usually use corn starch to marinate or create thickener. Personally I use potato starch as thickening agent and to marinate meat with corn starch.

The ratio of the power and water depends on the thickness required. Where thickener is concerned, I use a ratio of 1:3 for the powder to water, and a ratio of 1:1 for batter.

芡汁的質地

芡汁質地一般可分為流芡、糊芡、包芡及湯芡。流芡是一種稀薄的芡汁，能在碟中流動；糊芡則如其名呈糊狀，質感較濃稠。包芡的芡汁緊緊掛着食物表面，如咕嚕肉的芡。湯芡則呈稀薄及透明，多用於湯羹或燴菜中。

另外亦有分為湯汁芡、醬汁芡及西式芡三大類。

上湯芡的材料有生粉和上湯，上湯一般用排骨、老雞、金華火腿等經過數小時熬製而成，一般用於湯羹或紅燒菜式中如花膠鮑魚。

醬汁芡則主要在生粉水以外再加任何醬汁作調味，可濃稠可稀薄；如糖醋芡黑椒芡……質地較濃稠，可緊緊地裹着食材。

西式芡是加入了西餐元素的芡汁，以生粉混入花奶或茄汁等……如鴛鴦炒飯中的白汁芡及茄汁芡。不過亦有大廚做西式芡時會以麵粉再加入鮮奶和上湯做成西式芡。

Texture of thickener

The thickener texture is divided into liquid thickener, paste thickener, coating thickener and soup thickener. Liquid thickener is a thin thickener that flows through the dish; the paste thickener is thick in texture; the coating thickener coats the ingredients such as the thickener for sweet and sour pork; soup thickener is usually thin and transparent and is often used in soup or braised dishes.

It is also divided into 3 main categories : stock thickener, sauce thickener and western style thickener.

Stock thickener is made from potato starch and stock. Stock is usually made from hours of simmering pork rib, aged chicken and Jinhua ham. This thickener is generally used in thick soup or braised dishes such as fish maw and abalone.

Sauce thickener is made from potato starch solution added with other seasonings. It can be either thick or thin. Two examples are sweet and sour thickener or black pepper thickener. The texture is quite thick and able to fully coat on the ingredients.

Western style thickener is thickener added with some elements of western cuisine, such as potato starch mixed with evaporated milk or tomato sauce, the white sauce and tomato sauce used in Duo-styled fried rice. Some chefs would make a western thickener by mixing flour with evaporated milk and stock.

勾芡的方法

勾芡的方法原來亦很講究，一般為拌及澆。

拌是最常見的方法，此方法亦有兩種：一種是把芡汁加進接近完成的菜式中，快速拌抄後上碟，另一種是將芡汁下鑊煮至濃稠，再放入煮熟的食材拌勻，如炸好的海鮮或肉類如咕嚕肉之類。

澆則是將已完成的食材盛上碟，另起鑊將芡汁煮好後直接澆在菜式上，如翠塘豆腐的玻璃芡……以作最後調味外，亦為菜式增添光澤感，令賣相更佳。

芡汁遇上大火會立即結塊，宜於下芡後拌勻重新開火調校至合適的濃稠度，若繼續烹煮，芡汁會由稠身變回稀薄，整個芡汁會不能成形，可能要重新再煮。

Adding thickener

There are two ways to add the thickener to dishes: mixing or drizzling.

Mixing is the most common method and it can be done in two ways: one is to add thickener to a dish when it is almost done, stir well and dish up; second is to cook thickener in pan until texture thickens before mixing it with cooked dish such as deep-fried seafood or sweet and sour pork.

Whereas drizzling is to cook the thickener in pan and drizzle it over the cooked dish, such as the transparent thickener for tofu vegetable dish. Aside from enhancing the taste of the dish, it also adds a nice gloss to the presentation.

Thickener becomes lumpy when come into contact with high heat. After adding the thickener to the dish, it is advisable to turn on the heat to adjust to desired thickness. If it is over-cooked, the thickener will turn from thick to thin. The texture will be totally off and you may need to make a new thickener.

8 自製剁椒醬

剁椒醬有機會可以在雜貨店買到，但其實自己做亦十分簡單。

材料

朝天椒	500 克	鹽	40 克
大蒜	200 克	白酒	50 毫升
生薑	30 克	（二鍋頭或濃度50 以上的白酒）	
白糖	30 克		

做法

1) 將指天椒摘去根蒂後洗淨瀝乾水份，或用乾布印乾所有水份。大蒜去衣、生薑去皮後切小塊。

2) 指天椒放在乾淨無水無油的砧板上剁碎後（或用食物處理器攪碎），再放入一個乾淨的容器裏。
3) 把大蒜和生薑剁成薑蒜泥並加入剁椒內，加入餘下的調味料攪勻。
4) 用保鮮紙蓋上並蓋好蓋子，室溫下發酵半天後移入冰箱冷藏 7 天後即可食用。

貼士

請確保整個製作過程所有食材及工具滴水不沾。

Homemade Chopped Chilli Sauce

Though this chopped chilli sauce can be bought from the grocery stores, it is not that difficult to make it at home.

Ingredients

500g fresh red chilli

200g garlic

30g ginger

30g sugar

40g salt

50ml Chinese white wine (Erguotou or other white wine with alcohol volume more than 50%)

Steps

1) Remove the stems of chilli. Rinse and drain or use a dry cloth to pat dry the chillis. Peel the garlic and ginger, cut both into small pieces.
2) Finely chop the chilli on a dry the chopping board free of oil (or puree with food processor). Transfer to a clean container.
3) Mince the garlic and ginger and combine with the chopped chilli. Add the remaining seasoning ingredients. Mix well.
4) Cover with the cling wrap. Tighten the lid. Let ferment in room temperature for half a day. Chill in fridge for 7 days. Serve.

PS

All ingredients and tools must be completely dry and free of moisture.

9 調味料的份量指南

大家經常不知道調味料應該下多少才是，這裏有一個指南可供大家參考。最實際當然還是試味啦！（5 克約 1 茶匙，1 斤水大約是 600 毫升）

鹽	每 1 斤水加 7 克
雞精	每 1 斤水加 6 克
冰糖	使食材亮色，同時增加菜式回味，每 1 斤水加 2 克
料理酒	去腥，每 1 斤水加 5-8 克
白酒	異味較重的可用白酒去腥，正常每 1 斤水加 5 克
花雕酒	去腥增香，每 1 斤水加 6 克

Guide to the amount of seasonings

Most people are not too sure of how much seasoning is to be added. This guide will act as reference. The most practical way is to taste it personally. (1 teaspoon is about 5g, 1 catty of water is 600ml)

Salt	7 grams per 1 catty of water
Chicken bouillon	6 grams per 1 catty of water
Rock sugar	it can brighten up the colouring of the dish while enhancing the aftertaste; add 2 grams per 1 catty of water
Cooking wine	it is used to remove the fishy taste; add 5-8 grams per 1 catty of water.
Chinese white wine	it is used to reduce the strong odour of certain ingredients; generally to add 5 grams per 1 catty of water.
Huadiao wine	it is used to remove fishy taste and enhance the fragrance, add 6 grams per 1 catty of water

10 如何用香料

要做好滷菜，先要對各種香料有所了解，和其功效、作用以及使用方法。否則你的滷菜味道出現偏差也找不到原因。今天我們來簡單了解一下我們常用到的 28 種香料特色。

Using spices

To make a good brine dish, you must first understand the various spices, their efficacy, function and use. Otherwise you may not know why you just can't get the brine food to taste right. We will now take a brief look at the features of the dozens of spices.

主料介紹

八角、桂皮、山奈、丁香、草果、小茴香、白扣、川砂仁、香葉、白芷、甘草、陳皮、薑、花椒、辣椒。

Introduction to main spices

Star anise, cinnamon, safflower, clove, Cao Guo, fennel, cardamom, Chuan Sha Ren, basil, Bai Zhi, licorice, dried tangerine peel, ginger, Sichuan pepper, chilli.

輔料介紹

香果、華撥、紅寇、靈香草、益智仁、山楂、草蔻、五加皮、陽春砂仁、雲木香、良薑、甘松、排草、紫草、香茅草、千里香、香菜籽、紫蘇、桔實、桂枝、孜然、毛桃、川芎、當歸、黃芪、山藥、薄荷等……

Introduction to accessorized spices

Fragrant fruit, Bibo long pepper, Fructos galangal, Ling Xiang Cao, Yi Zhi, hawthorn, Cao Kou, Wu Jia Pi, Yangchun Sha Ren, Yunmuxiang, Liang Jiang, Gan Song, Pai Cao, comfrey, citronella, thyme , coriander seeds, perilla, Chinese citrus, cassia twig, cumin, peach, Chuan Xiong, angelica Dang Gui, Huang Qi, Chinese yam, mint, etc...

五香味的滷製，多以八角、桂皮為主，大概佔 35%。麻辣味的則以花椒、辣椒為主料，亦佔約 35%。其他 20 餘種去腥除異、脱臭增香、抗菌防腐香料等約佔 65%。如果按肉類特性區別的話，豬肉類異味較小，多以八角、桂皮為主。鴨肉異味較重，則多以白芷為主。所以，在配製滷料前，我們要根據各種肉類的味道特性來搭配香料比例。

Star anise and cinnamon make up 35% of the five spice brine. Those of "Mala" spicy numbing flavoured brines use Sichuan pepper and chilli as main ingredients, about 35%. The remaining 65% are made up of more than 20 spices for removing fishy or unpleasant odours, enhancing fragrance, with anti-bacterial and preservative properties. Different spices are used according to types of meat. Pork has less unpleasant odour, star anise and cinnamon are the main ingredients. Duck meat has stronger odour, Bai Zhi is the main spice. Therefore, before the brining, we need to decide on the spices and their ratio according to the taste and characteristics of the types of meat.

五香滷水包份量參考

八角20克、桂皮15克、草果10克、山奈10克、丁香3克、小茴香20克、白芷10克、白扣10克、草寇10克、陳皮5克、甘草10克、香果20克、當歸10、香葉10克、良薑10克、花椒20克、另加乾香菇30克。

這個份量太少，最好增加倍數再平均分勻多幾包。記得把材料輕微打碎。

新起滷水可以用1包，如滷得少，可在香料下鍋煮15分鐘聞到香味濃郁即撈出，再留待下次重用；如果滷得多，就讓香料包一直在滷鍋裏和肉一起滷。

Five spice brine pack component reference

20 g star anise, 15 g cinnamon, 10 g Cao Guo, 10 g Shan Nai, 3 g clove, 20 g fennel, 10 g Bai Zhi, 10 g cardamom, 10 g Cao Kou, 5 g dried tangerine peel, 10 g licorice, 20 g fragrant fruit Xiang Guo, 10g Dang Gui Angelica, 10 g fragrant leaf Xiang Ye, 10 g Liang Jiang, 20 g Sichuan pepper, plus 30 g dried Chinese mushroom.

The portions are at the low end. It is advisable to multiply the portions and make a few more packs by distributing the spices evenly. Lightly smash the spices before use. One pack is sufficient for new brine. While making brine foods of smaller quantity, cook for 15 minutes and remove the spice pack when the aroma is getting stronger. The spice pack can be reused. If the brine food quantity is more, just leave the spice pack together with the meat in the pot.

11 認識28種香料 Understanding 28 spices

1. 桂皮 Cinnamon

桂皮不管是在四川、廣西、山東等產地質量都不錯，有煙桂、肉桂、油桂、桂枝、紫桂、香桂等。好的桂皮厚實，乾脆，個大。如果你用手稍微掰一小塊下來，以味道濃厚沒有霉味的最好。

Good quality Cinnamon is produced in Sichuan, Guangxi, Shandong. They are divided into Yan Gui,

Rou Gui, You Gui, Guizhi cinnamon twig, Zi Gui, Xiang Gui and so on. Good cinnamon is thick, crisp and big. Break off a small piece to take a whiff, a strong but not musty smell indicates good quality.

2. 小茴香 Fennel seed

小茴香和八角的味道很夾，搭配在一起味道濃厚。小茴香能除去肉中臭氣，使之重新添香的作用，故曰「茴香」。小茴香如果買的不好，很容易有一股發霉的味道。選購時選綠色、顆粒大小勻稱、乾燥的。有些小茴香因存放時間太久而變黃，香味亦會流失。

The taste of fennel seed goes well together with star anise. Their combined flavour is rich and strong. Fennel seed can get rid of meat oadour and give it a pleasant aroma. Poor quality fennel seeds easily become musty. Pick the ones with green colour, dry and even in size.

3. 八角 Star anise

屬茴香料，產於福建、廣西、貴州、雲南、廣東、台灣、浙江等地。在滷肉滷水中可增香、去膩、去肉腥味的作用。八角不能選硫磺燻過的，味道不好而且對身體有害。買八角，要選個大均勻、香氣濃郁的。四川、廣西等的更好。

It belongs to the Illiciaceae family and is produced in Fujian, Guangxi, Guizhou, Yunnan, Guangdong, Taiwan, Zhejiang and other places. In the braised pork brine, it is used to enhance aroma, remove greasy taste and meaty odour. Do not pick star anise that is sulphur-treated, it tastes bad and is harmful to the body. Get those that are even and large with intense aroma. Better yet, get those produced in Sichuan and Guangxi.

4. 木香 Mu Xiang

原產於印度、緬甸、巴基斯坦。現分佈在四川，雲南，西藏等地。分類有川木香、雲木香、廣木香。特點是芳香氣濃味厚，在滷水中可增香去膩的作用。由於中藥味濃而苦，所以在滷水中不宜加多，否則便會太苦。可用溫水浸泡10分鐘左右，去除苦味也避免藥味過重後才用。採購時選個大、香味濃厚、整個的較好；切片的香味容易流失。

Native to India, Myanmar, Pakistan, it is now produced in Sichuan, Yunnan and Tibet. There are 3 classifications: Chuan Mu Xiang, Yun Mu Xiang and Guang Mu Xiang. It is characterized by a strong aroma with the effect to enhance fragrance and reduce greasiness in the brine. Due to its intense and bitter herbal taste, it is not advisable to add too much to the brine. Soak it in warm water for about 10 minutes to reduce the bitter and herbal taste before use. Buy those that are sold whole, large sized and with strong aroma; the sliced ones will lose the aroma quickly.

5. 草果 Cao Guo

主要出產於廣西、雲南等地，草果用來烹調菜餚，可去腥除羶，增加菜餚味道功效，煮滷水及牛肉常用。用時建議敲碎，只用殼除去籽，否則有苦怪味。

Mainly produced in Guangxi, Yunnan and other places, Cao Guo is used in cooking to enhance the taste of food. It is often used in brine and beef dishes. Break the fruit shell to remove the seed and use only the shell. Otherwise it will have a bitter and weird taste.

6. 山奈 Shan Nai

別名為沙薑、三柰、山辣。主要產於廣東、廣西、雲南、貴州、台灣等地。選用乾身及大一點、芳香濃厚廣西產的更好。在滷水中可增辛香，去腥羶味。

The alias is Sha Jiang, San Nai and Shan Lia. Mainly produced in Guangdong, Guangxi, Yunnan, Guizhou, Taiwan and other places. Choose those with a dry and bigger body. Even better, get the ones with intense fragrance produced in Guang Xi. It is added to the brine to remove unpleasant fishy and gamey meat taste.

7. 當歸 Angelica Dang Gui

主產於甘肅岷縣、武都、漳縣、成縣、文縣、湖北、四川、雲南等地。雲南的稱為雲歸。甘肅的較好。滷菜中加入當歸，有提香，壓臭味的功效，但能加多了就會發苦，藥味過重，用水先泡 10 分鐘更好。選個大而完整，香氣濃厚，大小勻稱的。

Mainly produced in Gansu County, Wudu, Zhang County, Cheng County, Wen County, Hubei, Sichuan, Yunnan and other places. Those produced in Yunnan is called Yun Gui. The ones from Gansu are better in quality. Adding angelica to the brined dish has the effect of improving fragrance and reducing odour However, if too much is added, it will cause bitterness and an overpowering herbal taste. It is better to soak angelica in water for 10 minutes before use. Pick the ones with a large, even and whole body with an intense aroma.

8. 陳皮 Dried tangerine peel

陳皮主要產於廣東、福建、四川等地。性溫味苦、辛，在滷水中可與山楂、生薑、木香宜配。陳皮不僅可促進消化，增進食慾，還可以去腥增加滷菜美味。陳皮的甘苦味可以與其他味道互相協調，因此可以用於調製滷水改善味道，不但辟去魚肉的羶腥味，且使菜餚特別芳香。

Dried tangerine peel is mainly produced in Guangdong, Fujian and Sichuan and other places. It is warm in nature, bitter and spicy in taste, goes well with hawthorn, ginger and Mu Xiang in the brine. Dried tangerine peel not only promotes digestion, increases appetite, but also enhances the taste of brined food. The bitter taste of dried tangerine peel can coordinate well with other flavours. It is often used in the preparation of brine to remove the fishy taste and to make the dishes particularly aromatic.

9. 丁香 Clove

有公丁香，母丁香之分。挑選丁香以個大、粗壯、色紅棕、油性足、能沉於水中、香氣濃郁、無碎末者為佳。個小、香味淡、有碎末的不宜選購。在滷水中可增香增回味、增食慾。香味持久，並有防腐保鮮的作用。用於滷水中不宜多放，否則會悶頭，味重。平時放兩粒丁香在口中還可除口臭。

It is divided into “male” and “female” cloves. The good quality ones are those with a large, thick, red-brown body that is rich in oil, without broken pieces, able to sink in the water and has a rich aroma. Do not get those that are small, light scented and with broken pieces. In the brine, it can improve food flavour and increase one's appetite. The fragrance is long-lasting. It also has antiseptic and preservative effect. Do not add too much to the brine, otherwise the taste will be too overpowering. To remove bad breath, put two cloves in the mouth.

10. 砂仁 Sha Ren

有川砂仁，陽春砂仁。福建、廣東和雲南等產地。果實完整，厚實，氣味芳香可增加滷水香味，在滷肉中使用選川砂仁為佳。

There are Chuan Sha Ren, Yang Chun Sha Ren. Mainly produced in Fujian, Guangdong and Yunnan. The fruit is whole, thick and fragrant, able to increase the flavor of the brine. For brining of meat, it is better to use Chuan Sha Ren.

11. 白扣 Bai Kou

原產於柬埔寨和泰國；中國則產於廣東、海南島、雲南和廣西等地。在滷水中作調味料，可去異味，增香。用於配製各種滷水以及滷豬肉、鴨、燒雞等……選天然白的才更健康，純白色的是經過人工漂白的並不太好。

Originally produced in Cambodia and Thailand. In China, it is produced in Guangdong, Hainan Island, Yunnan and Guangxi. As a seasoning in the brine, it can remove odours and enhance fragrance. For the preparation of a variety of brines and brined pork, duck, roast chicken etc. Use those with a natural white colour. Do not use the ones that are overly white and have been artificially bleached.

12. 草豆蔻 Cao Dou Kou

又名草蔻、草蔻仁、老蔻、老扣、又名土砂仁、假砂仁。主產於廣西、雲南等地。草豆蔻在滷水中有去除羶味、異味、怪味，增強特殊香味的作用。選用飽滿、較光澤、氣味芳香的更適合滷肉特點。

Also known as Cao Kou, Cao Kou Ren, Lao Kou, Tu Sha Ren or Jia Sha Ren. Mainly produced in Guangxi and Yunnan. It has the effect of removing gamey taste, odour, strange smell and enhancing the special aroma in the brine. Get those with a plump, shiny body and strong fragrance to better bring out the brined meat flavour.

13. 千里香 Qian Li Xiang

又稱七里香、萬里香、九秋香、九樹香、過山香、黃金桂等……主要產於廣東、福建、海南、廣西、湖南及貴州等地。在滷水中作調味料，可去異味，增香辛。也可用於滷牛肉、羊肉之等。

Also known as Qi Li Xiang, Wan Li Xiang, Jiu Qiu Xiang, Jiu Shu Xiang, Guo Shan Xiang, Huang Jin Gui, etc.... mainly produced in Guangdong, Fujian, Hainan, Guangxi, Hunan and Guizhou. As a seasoning in the brine, it can remove odour, enhance fragrance and spiciness. It can also be used to marinate meat such as beef and lamb.

14. 甘草 Licorice

分佈於新疆、內蒙古、寧夏、甘肅、山西朔州，以野生為主。在滷水中可以融合各味，複合味調味香料，採購時選新疆的甜味大，完整大片最佳。

It is grown in Xinjiang, Inner Mongolia, Ningxia, Gansu, Shanxi and Shuozhou, mainly in the wild. When used in the brine, it blends well with various flavours, seasonings and spices. The best quality ones are from Xinjiang with a strong sweetness and big intact slices.

15. 白芷 Bai Zhi

又稱香白芷，主產於四川、河北、安徽亳州、杭州、四川、杭州等地。選大個乾燥的川白芷根更佳。白芷片不好，香味易流失、也容易生蟲。白芷在滷水中可增香、去腥、去油膩、增進食慾。與白扣、砂仁配比組合為佳。

Also known as Xiang Bai Zhi, it is mainly produced in Sichuan, Hebei, Anhui, Bozhou, Hangzhou, Sichuan, Hangzhou. Chuan Bai Zhi root that is big and dry is of good quality. Sliced Bai Zhi is poorer in quality. The aroma is easily lost and it is easily damaged by worms. When used in brine, Bai Zhi can help enhance aroma, reduce odour and greasiness, improve one's appetite. The result is even better when used together with Bai Kou and Sha Ren.

16. 芫荽籽 Coriander seed

芫荽籽在滷水中可去腥提味，增厚味。但在滷水中不可多加，否則會發苦，藥味過重。

Coriander seed can be used in brine to enhance the taste, remove odour and thicken the flavour. Do not use too much. Otherwise it will become bitterish and the herbal taste will be too heavy.

17. 香茅草 Lemongrass

主要產於遼寧、河北、山東、河南、安徽、江蘇、浙江、江西、湖北、湖南、廣西、泰國等地。在滷水中的作用包括增加芳香、提香、飄香、並有抗菌防腐的功效，採購時選用芳香味濃，廣西、雲南、泰國的較好。

Mainly produced in Liaoning, Hebei, Shandong, Henan, Anhui, Jiangsu, Zhejiang, Jiangxi, Hubei, Hunan, Guangxi, Thailand. When used in brine, it can enhance fragrance, bring out the aroma, and has anti-bacterial and anti-septic properties. The better quality ones are those with rich aroma from Guangxi, Yunnan and Thailand.

18. 檀香 Sandalwood

木材奇香，香氣醇厚，經久不散。原產於印度、印尼、澳洲和非洲，台灣、廣東也有引種栽培。在滷水中增香效果非常明顯，一般 50 斤滷水裏加入 10 至 15 克為佳，如果加得太多，很容易適得其反。國內檀香假貨較多，是不法商家用藥水泡而成。產自於印度、印尼一帶的檀香品貌上乘，值得選購。

Sandalwood is incredibly fragrant. The fragrance is mellow and long lasting. It is native to India, Indonesia, Australia and Africa, and is also cultivated in Taiwan and Guangdong. The aromatizing effect in brine is very obvious. Generally, it is recommended to add 10 to 15 grams in 50 kg of brine. If too much is added, it can be counterproductive. There are quite a lot of fake sandalwoods within China, produced by unscrupulous businessmen through soaking normal wood in chemicals. Sandalwood from India and Indonesia is of top quality and worth buying.

19. 桂丁 Gui Ding

強烈芳香，味辛甘。它在滷水中可以殺菌抗氧化的作用。

Strongly aromatic, spicy and sweet in taste. When used in brine, it has anti-bacterial and anti-oxidation properties.

20. 決明子 Jue Ming Zi

決明子味苦、甘、咸，在滷肉中使用可以讓滷菜更入味的作用。

Jue Ming Zi has a bitter, sweet and salty taste. It is used in brining to allow the meat better able to draw in the flavours.

21. 羅漢果 Luo Han Guo

羅漢果味甜，味食香料，在滷水中使用可去腥，增加滷肉色澤的作用。

It is sweet and delicious; used in brine to remove any unpleasant odour and enhance the colour of the brined meat.

22. 檸檬乾 Dried lemon

檸檬乾在滷水中有去腥，提味，增加菜香的作用。

Dried lemon has the effect of odour-removing and taste-enhancing in the brine to improve the flavour of the dishes.

23. 白芍 Bai Shao

味苦、酸，在滷水中有去腥的作用。

It tastes bitter and sour with a deodorizing effect in brine.

24. 薄荷 Mint

薄荷芳香調料，味辛，用於滷製品中增加滷水香味。

Mint is an aromatic seasoning with a spicy taste. It is used in brined food to enhance the aroma of the brine.

25. 紫蘇 Perilla

味道辛、香，味道很香，也可以用於牛羊肉增香去腥。

It has a spicy, fragrant taste which can be used to reduce the gamey taste and enhance the flavour of beef and mutton.

26. 黃芪 Huang Qi

主要產於內蒙古、山西、黑龍江等地，味甘甜，在滷水中可去腥的作用。

Mainly produced in Inner Mongolia, Shanxi and Heilongjiang. It has a sweet taste and is able to remove fishy odour when used in brining.

27. 孜然 Cumin

孜然這種香料，多用於燒烤，燒羊肉等等，但其實它用於滷水中，便可除腥增香，但是別加太多，否則適得其反。

Although this spice is mostly used for barbecued meat, roast lamb, etc., it can also be used in brine to remove strong meat odour and enhance the fragrance; but do not put in too much, otherwise it can be counterproductive.

28. 薑黃 Turmeric

主產於福建、湖北、台灣、雲南、陝西等地，味道辛辣，有輕微橙味，香味較為特別，在滷水中可增加金黃色的作用。

Mainly produced in Fujian, Hubei, Taiwan, Yunnan and Shaanxi, the taste is spicy with a light orange flavour. The fragrance is rather unique. It is able to add a golden sheen to the brine.

Bonnie's Favourite Kitchen Appliances

Bonnie 廚房用具分享

1

一個好的冰箱可令你的食材保存得更好更長久。何謂好的冰箱？便是可以把冰箱內的濕度及溫度保持在非常穩定的水平。冰箱通常設定在 2-6℃內，冰格則在 -12 至 -18℃。當你每次打開冰箱時，便會把內裏的溫度及濕度改變而破壞食材存鮮度。所以，請盡量不要常把它大開中門，開門後要儘快關上。而好的冰箱能在開完門後以最快速度把水平調節好，我一直用 Fisher & Paykel 的。它能把我的蔬菜輕易地儲存兩，三個星期，大蜆也可存活 4 天。極力推薦。

A good refrigerator can keep the ingredients fresher for longer. What is a good refrigerator? It is one which can keep the humidity and temperature in the refrigerator at a very stable level. The refrigerator is usually set at 2-6 °C and the freezer between -12 and -18 °C. Every time the refrigerator door is opened, the temperature and humidity within will change and the freshness of the ingredients will be affected. Therefore do not leave the refrigerator door open too often. Close the door quickly. A good refrigerator can regulate the temperature and humidity very quickly back to the right level whenever the door has been opened. I have been using Fisher & Paykel all this while. It keeps my vegetables fresh for two or three weeks and the big clams alive for four days. I highly recommend it.

高速煲是我近年新寵。它可把要 1 小時煮好的東西，只需要 10 分鐘便弄好。它的強項是可以把食材快速煮腍，我平時要燜 18 小時的乾鮑魚，它只需 2、3 小時。當然缺點是可能不夠入味，但只要你先燜一半時間，由它浸過夜讓其入味，之後再燜餘下時間，便解決了這個問題。煮老火湯也是，4 小時的老火湯我現在煮 1 小時後，減壓後再中火滾半小時，便跟我平時煲的老火湯一樣味道了。有很多人對高速煲有些恐懼，只要你不要貪平，選些有國際標準的便可。而有 6 個排氣口，德國製造的 Fisher 一直是我的首選。

Pressure cooker is my new favorite in recent years. It only takes about 10 minutes to cook something that normally requires 1 hour. Its strength lines in the ability to tenderize the ingredients quickly. The dried abalone that I used to cook for 18 hours now needs only 2 to 3 hours to get done. The downside is that the abalones might not have fully absorbed the flavours. To overcome this, braise the abalones for half of the time and let steep overnight before carrying on the braising. Same goes for the long-simmering soups. A soup that normally takes 4 hours to cook, I will now cook it for 1 hour, reduce the pressure and cook for half an hour with medium heat. The taste is exactly like that of the long simmering soup. Some people may be somewhat fearful to use a pressure cooker. Just bear in mind to use one that is of international standard, not the cheap ones. With 6 air valves, the German brand Fisher is my first choice.

最近買了一個日本製的小型煙燻機，只需放入木碎、點火再開掣便成。機身輕巧，十分方便。從此告別舊式煙燻的麻煩了。

Recently I bought a small Japanese smoker. All that needs to be done is to add some wood chips, light the fire and turn on the switch. It is light and easy to handle. Here's goodbye to the cumbersome old-style smoking method.

4

我想向大家推介 Le Creuset 的易潔鑊，很多人説易潔鑊要保養得好，不能用大火又不能在高溫時下凍水洗。老實説我廚藝學校常常不聽話地粗用，但兩，三年仍然不黐底！塗層亦沒有脱落。

I will introduce the non-stick pans Le Creuset. Most people claim that to keep a non-stick pan in good condition, avoid cooking with high heat or rinsing a hot pan with cold water. To be honest, I have not been handling the pans that gently in my cooking classes but the non-stick layer is still in place and the ingredients still come off easily.

Kenwood 的廚師機一向是我廚房的好幫手，加上不同配件又有不同功能。打麵糰夠力，不容易燒摩打。又可以打醬汁、攪肉、做雪糕、做麵、切麵……是煮食必備之選。

Kenwood Kitchen Machine has always been a good helper in my kitchen. It comes with various accessories to perform different functions. Aside from being a powerful dough kneader without the engine getting over-heated, it can whip up sauces, blend meat, make ice cream, make and cut noodles... It is a must-have kitchen tool.

大家也不只一次見我推介極力法國 Cristel 五層不銹鋼中式鑊了，很夠火候又不易黐底，電磁爐及明火也合用。全不銹鋼就算給工人姐姐粗用也不怕，極力推介。

This is not the first time I highly recommend the French Cristel five-layer stainless steel Chinese wok. It has high thermal efficiency yet is stick resistant, can be used on induction cooker or open fire. The all stainless steel material can withstand rough use by domestic assistants. Highly recommended.

星級賞

頭獎（1 名）

Fisher & Paykel ActiveSmart™ 635 毫米雪櫃底部冷凍櫃帶製冰機飲水機 359 公升 E402BRXFDU4 (價值 HK$27,080)

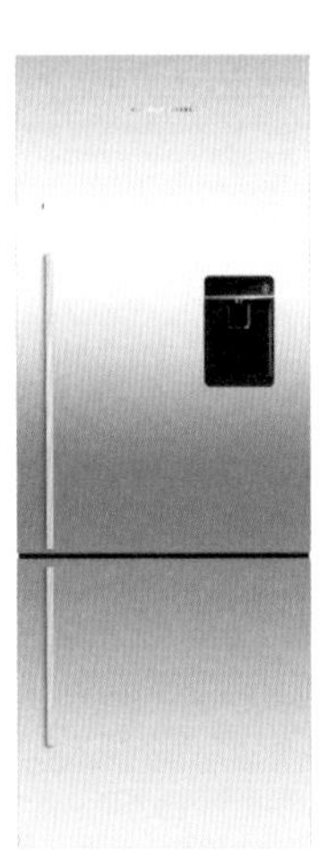

二獎（1 名）

Le Creuset

Flame 圓形琺瑯鑄鐵鍋 18 厘米 (價值 HK$2,368)

安慰獎（14 名）

Le Creuset

Kiwi 陶瓷心形小鍋子連蓋 (每個價值 HK$338)

參加辦法

- Step 1 讚好萬里機構及玻璃朱 Facebook 專頁。
- Step2 在《星級菜簡單煮》post 內 Tag 3 位朋友。
- Step 3 Inbox 我們提供資料，包括姓名、電郵、聯絡電話、購買本書的單據證明。
- Step 4 記緊保留購買本書單據，拍照再傳送給我們。
- Step 5 然後回答以下問題：你最想嘗試的是哪一個食譜？簡單分享原因。

*萬里機構 facebook: https://www.facebook.com/wanlibk/ *如有任何爭議，萬里機構出版有限公司保留最終決定權。
*頭獎獎品會由贊助商提供送貨服務。*二獎及安慰獎需自行領取獎品。*得獎者將獲個別通知。*截止日期：2019 年 9 月 30 日。

Simple Signature Dishes

作者 Author
玻璃朱 Bonnie Chu

策劃 Planner
謝妙華 Pheona Tse

編輯 Editor
李穎宜 Wing Li

美術設計 Designer
鍾啟善 Nora Chung

攝影 Photographer
Humblejim Ltd
梁細權 Sai Keung

化妝及髮型 Make-up & Hair Styling
Meegan@Meegan make-up studio
Rachel@Blissbridecouture

出版者 Publisher
萬里機構出版社有限公司 Wan Li Books Company Limited
香港鰂魚涌英皇道1065號 Room 1305, Eastern Centre, 1065 King's Road,
東達中心1305室 Quarry Bay, Hong Kong.
電話 Tel: 2564 7511
傳真 Fax: 2565 5539
電郵 Email: info@wanlibk.com
網址 Web Site: http://www.wanlibk.com
http://www.facebook.com/wanlibk

發行者 Distributor
香港聯合書刊物流有限公司 SUP Publishing Logistics (HK) Ltd.
香港新界大埔汀麗路36號 3/F., C&C Building, 36 Ting Lai Road,
中華商務印刷大廈3字樓 Tai Po, N.T., Hong Kong
電話 Tel: 2150 2100
傳真 Fax: 2407 3062
電郵 Email: info@suplogistics.com.hk

承印者 Printer
中華商務彩色印刷有限公司 C & C Offset Printing Co., Ltd.

出版日期 Publishing Date
二零一九年七月第一次印刷 First print in July 2019

Published in Hong Kong
ISBN 978-962-14-7028-7

特別鳴謝：
Le Creuset 提供部分餐具
Fisher & Paykel 提供拍攝場地